SINTERING TECHNIQUES OF MATERIALS

Edited by **Arunachalam Lakshmanan**

Sintering Techniques of Materials
http://dx.doi.org/10.5772/58496
Edited by Arunachalam Lakshmanan

Contributors

Ming Yan, Yaniv Gelbstein, Yedidia Haim, Sergei Kalabukhov, Vladimir Kasiyan, S Hartman, S. Rothe, Nahum Frage, Aimin Chang, Sergio Cava, Tiago Delbrücke, Rogério Gouvêa, Cristiane Raubach, Mário Moreira, Jose Varela, Elson Longo, José Jurado, Faili Cintia Tomsen Veiga, Vânia De Sousa, Iwona Sulima, Makoto Nanko, Khanh Quoc Dang, Uilame Umbelino Gomes, Barbara Staniewicz-Brudnik, Elżbieta Bączek, Grzegorz Skrabalak, Arunachalam Lakshmanan

Published by InTech

Janeza Trdine 9, 51000 Rijeka, Croatia

Publishing Process Manager Iva Simcic

Technical Editor InTech DTP team

Cover InTech Design team

First published March, 2015

Printed in Croatia

Additional hard copies can be obtained from orders@intechopen.com

Sintering Techniques of Materials, Edited by Arunachalam Lakshmanan

p. cm.

ISBN 978-953-51-2033-9

For EU product safety concerns:
IN TECH d.o.o., Prolaz Marije Krucifikse Kozulić 3, 51000 Rijeka, Croatia,
info@intechopen.com or visit our website at intechopen.com.

Meet the editor

Dr. Arunachalam Lakshmanan obtained his Ph.D on Thermoluminescence Dosimetry during his tenure as Scientist at BARC, Mumbai from 1972-90. He was heading the Radiation Dosimetry Section at IGCAR, Kalpakkam during 1990-2007. Presently he is a Research Professor in Physics at Saveetha Engineering College, Chennai. His research areas include recycling PTFE, synthesis of new luminescence phosphors for display, lighting, X-ray imaging and radiation dosimetry. Dr. Arunachalam Lakshmanan is currently working on the development of a third form (other than fission or fusion) of nuclear energy which is eternal and green. He has earlier worked on CERN, Geneva as a scientific Associate and as AvH Fellow at University of Wuppertal, Germany and as a visiting Professor at Postech, South Korea. He has written books and published several papers in international journals and is presently guiding students in their doctoral thesis research work on luminescence.

Contents

Preface

Sintering is the process of fusing together ceramic, metallic or plastic particles under heat and pressure, blending them to enhance their strength and other properties, as well as wear life. Most materials products for use in key industries such as electronics, nuclear and aerospace are fabricated by sintering. Lots of novel sintering technologies, such as spark plasma sintering, mechanical alloying combined with hot isostatic pressing, and microwave sintering, have been developed during the past twenty years. However, it is still crucial to have processing methods to fabricate dense components by techniques that can retain the particle size or designed composition distribution in the final microstructure.

The present book covers new sintering techniques on ceramic materials, metals and composites as well as reprocessed PTFE. This is a continuation of an earlier book published by InTech in 2012 on Sintering of Ceramics – New Emerging Techniques edited by me. The book covers theoretical as well as experimental aspects on Spark Plasma Sintered (SPS) Porous copper, development of cutting blades with high hardness and resistance to cracking and wear, increased microhardness of austenitic steel–TiB2 composites obtained with high pressure - high temperature sintering, Al_2O_3 porous body with cotton as the template and excellent thermal insulation with direct application for refractories as well as metal matrix composites added nanostructured tantalum carbide and an overview of different sintering techniques used in powder metallurgy. In the case of SPS, the combination of internal heating and external pressure provides the conditions for fast sintering. The mostly popular laser-based AM techniques, namely selective laser melting (SLM), electron beam melting (EBM), laser metal deposition (LMD) and selective laser sintering (SLS) for the fabrication of Ti-6Al-4V have been overviewed based on an analysis of over 100 individual studies. Finally recycling of PTFE scrap materials using ram extrusion and compression molding are described.

Arunachalam Lakshmanan
Saveetha Engineering College,
Thandalam, Chennai,
India

Section 1

Sintering of Ceramic Materials

Chapter 1

Pulsed Electric Current Sintering of Transparent Alumina Ceramics

Makoto Nanko and Khanh Quoc Dang

Additional information is available at the end of the chapter

http://dx.doi.org/10.5772/59170

1. Introduction

Aluminum oxide (Al_2O_3) commonly referred as to alumina is one of the most widely used as engineering oxide ceramics. From crystalline structure difference, there are many forms of Al_2O_3 (α, χ, η, δ, θ, γ and ϱ), with α-Al_2O_3 being thermodynamically the most stable form. An example of α phase of Al_2O_3 is corundum or sapphire [1]. In the present chapter, α-Al_2O_3 is discussed and described as Al_2O_3. With a high melting temperature, chemical stability, Al_2O_3 is leading to applications as high-temperature components, catalyst substrates and biomedical implants. Al_2O_3 has excellent optical transparency and along with additives such as chromium and titanium, it is important as a sodium lamp (sapphire), a gem stone (sapphire and ruby) and a laser host (ruby).

Usually, Al_2O_3 ceramics were produced by sintering Al_2O_3 powder, that is, polycrystalline Al_2O_3. Sintered polycrystalline Al_2O_3 ceramics were opaque because of light scattering by closed pores and grain boundaries. In order to fabricate transparent polycrystalline Al_2O_3, many sintering techniques have been studied such as hot-pressing (HP), hot isostatic pressing (HIP), microwave sintering and pulsed electric current sintering (PECS).

PECS is also known as spark plasma sintering (SPS) or plasma activated sintering (PAS). The sintering technique is the latest pressure-sintering process to consolidate advanced materials such as ceramics, metallic materials, composites, polymers, semiconductors and oxide superconductors, in which the powder is heated by the application of electric current under uniaxial pressure.

PECS is a promising sintering technique for producing transparent polycrystalline Al_2O_3. In the present chapter, progress in PECS for transparent Al_2O_3 was discussed as well as other oxides. Fundamentals of PECS were also discussed in the present chapter in order to understand PECS for transparent Al_2O_3.

2. Fundamentals on PECS

According to the open literatures [2, 3], the electric current activated/assisted sintering technology was pioneered by Duval d'Adrian in 1922 [4]. However, the first patent on pure direct current (DC) resistance sintering (RS) was proclaimed by Bloxam in 1906 [5, 6]. Thereafter, Taylor [7-9] developed the resistive sintering process consisting of capacitors, transformers and special switching devices. This process originated the electric discharge compaction (EDC) [10].

Inoue [11, 12] developed the first concept of the PECS technology in 1966. It introduced different electric current waveforms, i.e. low-frequency alternate current (AC), high-frequency unidirectional AC or pulsed DC. These sintering techniques were combined in one sintering process of electric-discharge sintering (EDS) [11], also known as spark sintering (SS). In SS process, a unidirectional pulsed DC or a unidirectional AC, is applied, then DC is eventually superimposed. This process led the development of current PECS technology, e.g. plasma activated sintering (PAS), spark plasma sintering (SPS), filed assisted sintering and plasma pressure compaction® (P^2C) [13].

In the late 1980s various companies started to manufacture PECS machines based on Inoue's patents. Since then, the number of the PECS applications has been extended further. In the early 1990s, Sumitomo Coal Mining Co. commercialized the new PECS apparatuses (2-20 kA DC pulse generators, 98-980 kN load cells) [14, 15]. The PECS process is schematically shown in Figure 1.

It simultaneously applies an electric current along with a uniaxial pressure in order to accelerate densification of powders with desired configuration [16]. The electric current delivered during PECS processes could in general assume different intensity and waveform which depend upon the power supply characteristics [2, 3, 14, 16].

The PECS process is characterized by the application of the pulsed electric current during sintering. The heating rate in the PECS process depends on the materials and shapes of the die/sample ensemble and on the electric power supply. Heating rates from 100 to 600 K/min can be obtained in the current PECS equipments. As a consequence, the PECS process can be in time ranges from a few ten seconds to minutes depending on the material and its size to be sintered, configuration and equipment capacity.

The temperature is measured either with a pyrometer focused on the surface of the graphite die or with the thermocouple inserted into the die. Usually, the measured temperature at the surface of the die (die temperature) is lower than that of the sample (sample temperature). The magnitude of this temperature difference depends on a number of factors such as thermal conductivity of the die and the sample, the heating rate used, the pressure used, how well the die is thermal insulated etc. [17]. The current and consequent temperature distributions within

http://dx.doi.org/10.5772/59170

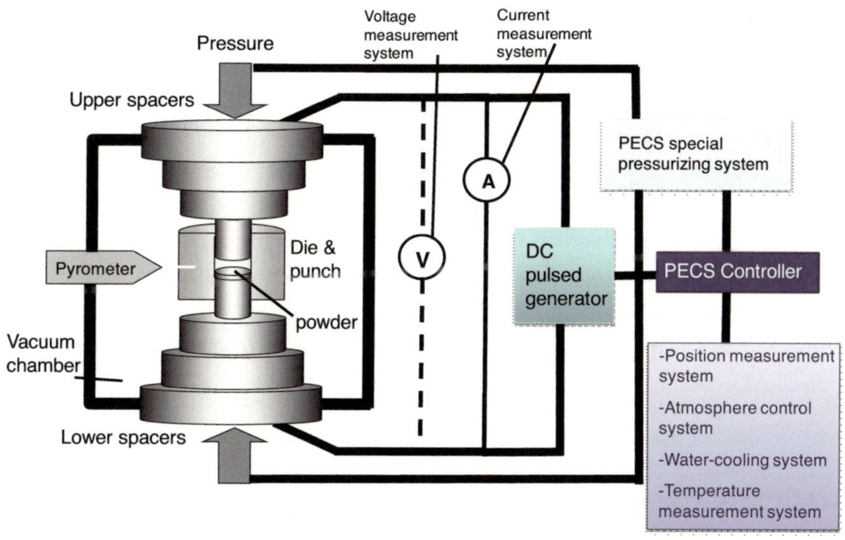

Figure 1. Schematic representation of the PECS process

the sample inside are very important to the homogeneity of density and grain size distribution of the product. Locally dense parts, at the beginning of current flow in particular, may result in locally overheating or even melting [16]. Experimental evidence of temperature distributions with different conductivity materials have been reported in [18-24]. It has been verified that the electrical properties of the sample influence significantly the temperature distributions inside the die as well as sample inside. Thus, in a nonconductive sample (i.e. Si_3N_4 and Al_2O_3), larger thermal gradients has been sometimes observed than in the case of a conductive one (i.e. Ti and Ni), indicating that the temperature distribution within the nonconductive sample is not as homogeneous as within a conductive sample.

Current understanding of the effects of pulse current waveform on compact density in a PECS process is still incomplete. The pulse current did not affect significantly the PECS of cast-iron powder [20] and Ni-20Cr powder [21]. In the PECS process of Al powder, densification behavior is independent of pulse frequency ranging from 300 Hz to 20 kHz [25]. The applied current, however, can significantly affect the growth of the product layer in chemical reaction between Mo and Si plates [26, 27]. With the PECS process, the pulse DC current affected the growth of Nb-C system, Mo_2C layer formed in Mo/C, Ti/C and Zr/C diffusion couples [28-30]. Inoue claimed that there was a frequency-dependent effect in his patent [31]. The densification rate of Fe and Ni based alloy processed by pulsed current was about 5% faster than by direct current [32].

However, the sintering mechanism of insulating oxides such as Al_2O_3 using the PECS method is still an on-going research area. Many papers on PECS of Al_2O_3 powder focused on densification and grain growth behavior by investigating effects of various parameters such as particle size, heating rate, sintering time, pressure and sintering temperature during the PECS process. Influences of the sintering parameters on densification and grain growth are not clear yet.

There are no reports about pulse current waveform effects on sintering behavior of Al_2O_3. The waveform of applied current is probably an important factor to the sintering process of Al_2O_3. An effect of two types of pulse current waveforms, inverter and pulsed DC, on sample temperature and densification of Al_2O_3 powder by using the PECS process has been clarified in [33-35]. The magnitude of the voltage peaks increased with an increase of the "OFF" time relative to the "ON" time for all of pulse power generator. Maximum voltage value of the inverter generator was higher than that of the pulsed DC generator. PECS with the inverter generator had higher sample temperature than that with the pulsed DC generator.

In PECS of Al_2O_3 powder, the electric current would be mostly applied to the punches and graphite die for heating up to sintering temperature. The average peak height of the 12/2 pulsed DC pattern is lower than that of the 2/6 pulsed DC pattern as well as lower than that of 40/10 inverter and 10/20 inverter pattern at the same die temperature. The inverter-type PECS had a higher voltage applied to the graphite die than the pulsed DC-type ones at the same die temperature. When the number of the "OFF" pulses increased as in the 10/20 inverter or the 2/6 pulsed DC pattern, the peak height of voltages of the "ON" pulses must have increased to keep the output power constant.

Temperature difference in Al_2O_3 sample is generated in PECS [33-35]. When PECS of Al_2O_3 sample with ϕ15 in diameter and 3 mm in thickness was conducted, temperature of sample outside was 20 - 30 K higher than that of the inside sample. The difference of inside/outside temperature using pulsed DC was approximately 10 K lower compared to the inside/outside temperature using the inverter. PECS with an inverter had a higher sample temperature than that with a pulsed DC power generator and it also higher than the die temperature. When the die temperature is increased, the temperature difference between the die surface and the sample also increases.

The sample temperature would be strongly affected by the applied current profile during the PECS process. The current flow should be strongly dependent on the characteristics of the different elements which compose the system (powder, punches, die) and, particularly, their electrical and thermal characteristics. For an insulating material, the applied current does not flow through the sample when a pulse current power is applied to the die-sample, but could only flow from one punch to the other punch via the die. The current forms a magnetic field in the near surface of punches and the die inside where is close to the sample surface, and this magnetic field affects current density [19, 33-37]. The highest current density should be located close to the sample surface as can be illustrated in Figure 2. The temperature distribution is closely related to the current distribution because the heat transfer is generated by the flow of current at the graphite die and the punches. Thus, during the PECS process, the Al_2O_3 powder must be sintered by the heat transferred from the die inside close to the sample surface and punches by means of heat conduction. Given that heat generation and transfer lead to a temperature distribution, temperature outside is higher than that inside the sample [33-35]. In the punch-compression direction, the die temperature is lower than the sample because the punches are in contact with water-cooled jacket and the die is cooled by radiation from the die outer surface.

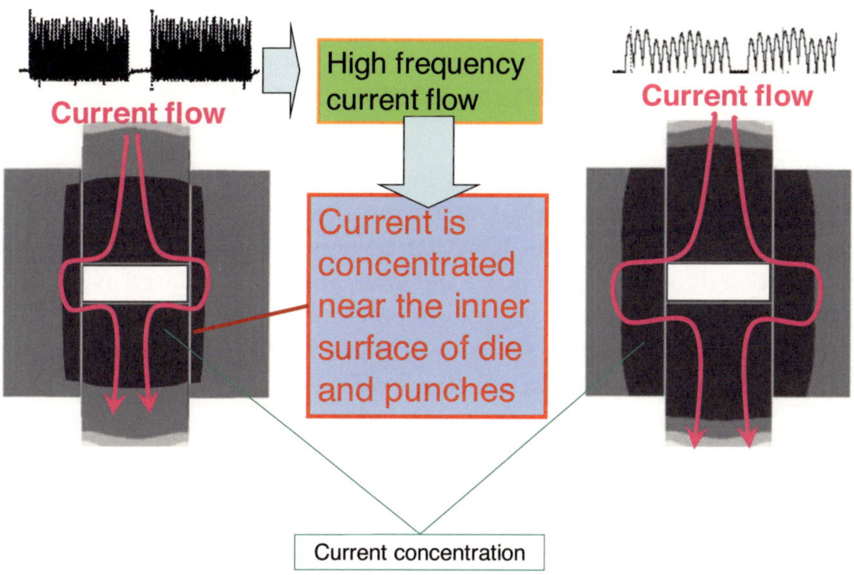

Figure 2. Current flows and distributions in the PECS die/punch/sample system for various pulse waveforms.

The ON/OFF pulse patterns and power generator frequency could also affect the sample temperature. In the case of inverter power waveform, high voltage with high frequency at long "OFF" time flows into die and punch gives higher heat transfer to heat sample than that of other pulse patterns. The difference in sample temperature using pulse current waveform inverter and pulsed DC could be explained by the skin effect as shown in Figure 2. In the punch/die/sample system, the back color area shows the current concentration during the PECS process. In both cases, the heat is generated only in the conductive die and the distribution of the heat generation does not change drastically with the electric conductivity of sample. Electric current distribution is the main cause of the temperature gradient between the sample and the external surface of the die, together with radiation heat from the die surface. When the high frequency current (inverter generator) is applied to punches and graphite die, the current density near the inner surface of the punches/die should be higher than that at its center. In contrast, at low frequency (pulsed DC generator), current density would be uniformly distributed across graphite die and punches. This difference suggests that higher temperature could be achieved due to higher applied voltages, and the required energy for heating a sample with an inverter is higher than that with pulsed DC generator [19, 33-37]. The relative density as a function of the outside/inside sample temperature was discussed in [33-37]. These results show a consistent relative density increase trend with an increase in the sample temperature, independent of the applied pulse current waveforms and ON/OFF patterns. It was also revealed that the average grain size increases with an increase in sample temperature even in different pulse current waveforms and ON/OFF patterns. Densification and grain growth were predominated by sample temperature. The pulse electric current waveform had effects on the sample temperature, but did not have direct influence on the densification, grain growth and homogeneity of the sample sintered by the PECS process.

3. Sintering of transparent polycrystalline alumina

Transparent polycrystalline Al_2O_3 has increasingly become the focus of recent investigations primarily because of their unique combination of properties. Single crystals of Al_2O_3 are highly transparent in visible and IR region. However polycrystalline Al_2O_3 ceramics are usually opaque because of light scattering of pores and grain boundaries as well as impurities. High density is the most important factor to produce polycrystalline transparent ceramics, as well as grain size. Because of the high efficiency of pores for light scattering, transparency in polycrystalline materials requires extremely low level in porosity, less than 0.01 vol.%. Samples with such low porosity could only be produced under proper sintering conditions involving high temperatures and long sintering time. Residual porosity is much more important than grain boundaries for obtaining the transparency, even in crystallographically anisotropic materials in optical properties. The scattering efficiency for spherical pores, however, decreases dramatically when the pore size in the nanometric range could be achieved [38-40]. It is believed that nanostructured polycrystalline materials would possess higher transparency than ones with the micrometric grain size range. The sintering process at high temperature causes extensive grain growth and then seriously degrades the mechanical properties of the material. What is more important, the higher/bigger grain size larger than 410 μm leads to significant light scattering coming from the birefringence of coarse Al_2O_3 grains [41].

After Coble developed transparent polycrystalline Al_2O_3 [42], many studies for producing transparent polycrystalline Al_2O_3 by sintering techniques were reported [40, 41, 43-82]. Recently, fine-grained transparent polycrystalline Al_2O_3 has attracted much attention due to its superior mechanical and optical properties. This material is prepared by sintering using HP and HIP at low temperature ranging from 1150 to 1400°C. The formation of nanostructure (< 1 μm) results in a significant improvement in both the mechanical strength and the optical transparency. It is reported that the mechanical strength of the fine-grained transparent Al_2O_3 is reached up to 400 - 600 MPa together with a high in-line transmission up to 60 % for visible light [41, 45]. Thus far, the addition of small amount of MgO is known to suppress normal and abnormal grain growth. The MgO concentration needed to inhibit abnormal grain growth depends on other impurities, CaO, SiO_2 etc.. [42, 46, 47, 70, 72]. Coble opened a new chapter that positive effect of 250 ppm MgO addition in sintering of Al_2O_3 is accompanied by dissolution into Al_2O_3 and excess MgO beyond its solid solubility limit exists as non-stoichiometric $MgAl_2O_4$ spinel at the grain boundaries of Al_2O_3 [42]. Hence, MgO strongly segregates into Al_2O_3 grain boundaries and produces a solute drag effect. The resultant microstructure is finer in grain size with higher final density. The transparent MgO doped Al_2O_3 ceramics was sintered to full density and had an in-line transmission of 40-50 % between 400 and 600 nm of the wavelength.

On the other hand, HPed Al_2O_3 yielding better transparency than pressureless-sintered samples was reported long back [38, 48-50]. Those reports showed that the increase of the transparent Al_2O_3 with a much smaller grain size of 1 μm could be obtained by a continuous hot-pressing process at 1400°C under pressures of 120 MPa in different atmosphere.

The major contribution to the less transparency for undoped polycrystalline Al_2O_3 originates from scattering caused by the remaining pores, and the difference between translucent and transparent Al_2O_3 could come entirely from the difference in pore-size and its distribution. Until quite recently, HIP is the most widely used technique for developing transparent alumina as it eliminates residual porosity and prevents grain growth leading to high transmission. For post-HIP treated samples, results for undoped [52], single doped (Mg^{2+}, Ti^{4+}) [39, 45, 46] transparent polycrystalline Al_2O_3 have been reported so far. This method gave fairly reproducible in-line transmittance between different groups with values up to 65 %.

Recently, some new sintering techniques such as microwave sintering [65] have also been studied for transparent crystalline Al_2O_3. Pressure-sintering such as HP and HIP is usually expensive in process cost. Microwave sintering is expected to realize homogeneous heating of the whole of ceramic sample.

4. Transparent Polycrystalline Al_2O_3 Produced by using PECS

Many reports on PECS for sintering transparent Al_2O_3 have been published as well as other transparent polycrystalline oxides. As process technology for ceramic powder is progressed, oxide ceramic powders with fine grain size and less agglomeration have been developed. Transparent polycrystalline Al_2O_3 with fine grains have been able to be prepared with such advanced oxide powder by using PECS.

Recently transparent oxide ceramics with fine grains such as 300 nm have been reported with different the PECS techniques as well as Al_2O_3. Munir and his colleagues promote PECS with ultra-high pressure such as 500 MPa [82]. High-pressure PECS is effective for preparing highly transparent polycrystalline Al_2O_3 [83], and also Y_2O_3-doped ZrO_2 [82] and Y_2O_3 [84]. High-pressure PECS is very useful for eliminate closed pores. However sample size is likely limited in high-pressure PECS.

Kim et al. proposed slow-heating PECS for densifying Al_2O_3 with less grain growth [57]. PECS with slow healing rate is available for not only Al_2O_3 but also $MgAl_2O_4$ [85]. Kim studied kinetics of densification and grain growth with stress rate in the point of view on "dynamic grain growth" [86]. He mentioned that slow stress rate in PECS is preferred in order to densification of Al_2O_3 with less grain growth. On the other hand, Makino and his colleagues reported that transparent polycrystalline Al_2O_3 can successfully obtained by PECS with fast heating rate such as 200 K/min [73]. However transparency of the sample with fast heating rate was not good in homogeneity. In order to densify Al_2O_3 without significant grain growth, influences of heating rate is still in discussion.

Goto and his colleagues reported PECS of transparent Lu_2O_3 with two-step pressure profile [87]. Lu_2O_3 is one of the candidates on laser host materials for high-power and ultra-short pulse lasers. However it is difficult for densification by conventional sintering. Taking account of advanced studies on transparent oxides given by Kim and Goto, a sintering profile is very important even in a process of PECS.

Thus PECS provides transparent polycrystalline oxides. Besides the oxides described here, there are many examples of transparent oxides sintered by using PECS. Table 1 shows a variety of transparent polycrystalline oxides prepared by using PECS.

Materials	Dopants	Remarks	Ref.
Al_2O_3	As-received	Slow-heating	[57]
	Cr_2O_3-doped		[76]
	As-received	High-pressure	[88]
	MgO, Y_2O_3 and La_2O_3-doped		[56]
	Cr_2O_3-doped	Slow-heating	[77]
	ZrO_2, La_2O_3 and MgO- doping		[78]
	MgO-doped	High-pressure	[83]
	As-received	Fast-heating	[73]
	As-received	Slow-heating	[80]
	La_2O_3-Doped	High-pressure	[81]
	As-received	Two-step temperature	[74, 75]
$MgAl_2O_4$	Undoped & LiF-doped	Two-step Pressure & Temperature	[89]
	As-received	Slow-heating	[85]
	As-received		[90]
	As-received		[91]
	As-received	Slow-heating	[92]
	Lab-made		[93]
Y_2O_3-Doped ZrO_2	As-received	High-pressure	[82]
	As-received		[94]
	As-received		[95]
	Lab-made		[96]
Y_2O_3	As-received, undoped	High-pssure	[97]

Materials	Dopants	Remarks	Ref.
	As-received, undoped		[98]
Lu_2O_3	As-received, undoped		[99]
	As-received, undoped	Two-step Pressure	[87]
	Lab-made, Yb_2O_3-doped		[100]
MgO	As-received		[101]
	Lab-made, Undoped & CaO-doped	High-pressure	[102]
$Lu_2Ti_2O_7$	Lab-made, undoped	Two-step Pressure	[103]
Lu_3NdO_7	Lab-made, undoped	Two-step Pressure	[104]
$La_2Zr_2O_7$	Lab-made, undoped	Two-step Pressure	[105]
β-$Ca_3(PO_4)_2$	undoped		[106]
YAG	undoped		[107]
	undoped &LiF-doped		[108]
	Lab-made, undoped		[109]
Oxyapatite	Lab-made, undoped		[110]
	Lab-made, undoped	High-pssure	[111]
Mullite	Lab-made, undoped		[112]

Table 1. Transparent polycrystalline oxides produced by PECS

5. Two-step PECS for transparent polycrystalline alumina

The authors study PECS with two-step temperature profile, that is, two-step PECS (referred as to TS-PECS), in order to fabricate transparent oxide ceramics with fine grains [74, 75]. Figure 3 shows the sintering profile of TS-PECS with other PECS techniques. TS-PECS can provide well-transparent oxides with shorter sintering period in comparison with slow-heating PECS.

Figure 4 shows appearance, fracture surface and density of polycrystalline Al_2O_3 prepared by using TS-PECS with 1st different temperature for 60 min and 1200°C for 20 min under 100 MPa in vacuum. A sample prepared by slow-heating PECS at 1200°C is shown for comparison. Importance of the 1st step temperature can be understood in Figure 4. The sample sintered at 1000°C in the 1st step has high transparency and less grain growth. The meaning of the 1st step is densification without significant grain growth. Sintering at 1000°C can provide densification without grain growth, however, full densification cannot be achieved. In order to reach to the full densification of the sample, the 2nd step with higher sintering temperature is necessary.

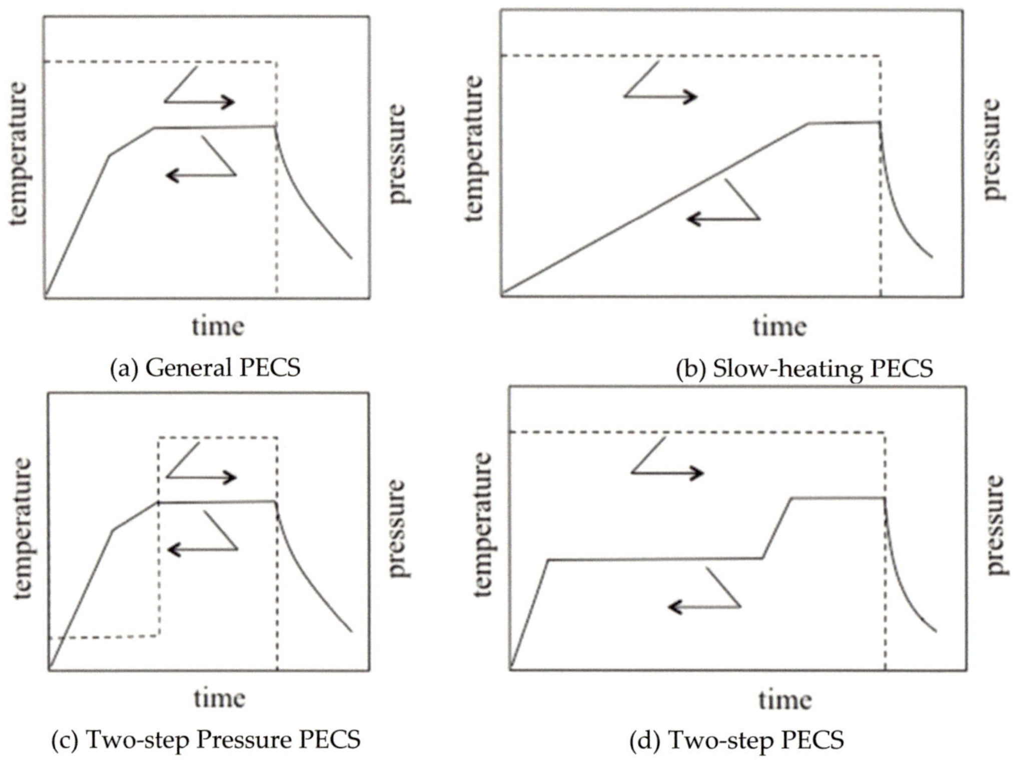

Figure 3. Schematic diagrams on sintering profiles of TS-PECS with other PECS techniques.

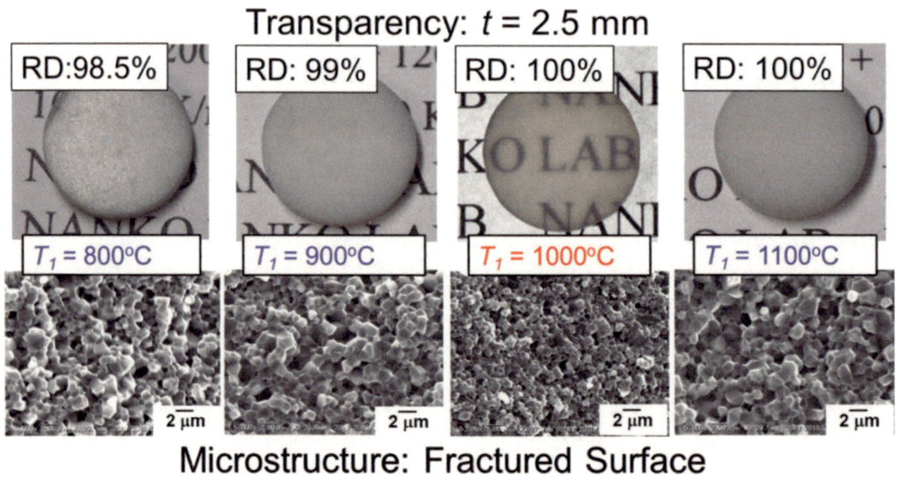

Figure 4. Appearance, fracture surface and density of polycrystalline Al_2O_3 prepared by using TS-PECS with 1st different temperature for 60 min and 1200°C for 20 min under 100 MPa in vacuum.

TS-PECS is also useful for other transparent oxides such as $MgAl_2O_4$. This polycrystalline oxide has better transparency because of isotropic crystal structure. Figure 5 shows appearance of polycrystalline $MgAl_2O_4$ produced by using TS-PECS. Even regular PECS such as 1300°C for 20 min with 100 K /min can provide transparent $MgAl_2O_4$ with fine grain size. However TS-PECS can increase transparency of the sintered sample.

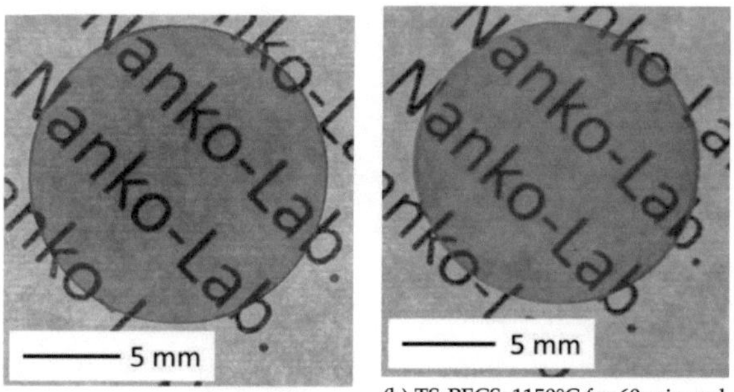

(a) PECS: 1300°C for 20 min

(b) TS-PECS: 1150°C for 60 min and 1350°C for 20 min

Figure 5. Appearance of polycrystalline $MgAl_2O_4$ produced by using (a) PECS and (b) TS-PECS under 100 MPa in vacuum.

Table 2 shows mechanical properties of transparent Al_2O_3 of TS-PECS. Bending strength of the samples is approximately 400 MPa, which is comparable with any commercial opaque Al_2O_3. This is caused by the existence of macroscopic defects as large as a few tens micrometers. Figure 6 shows an optical microscopic image of the inside of the transparent Al_2O_3 prepared by TS-PECS. Many black dots are observed in the sample. Figure 7 represents a scanning electron microscopic image of the cross-section of a black dot in transparent Al_2O_3 prepared by TS-PECS. Size of the black dot in Al_2O_3 is approximately 50 μm in diameter. The black dot is pores although the surrounding is fully densified. The microstructure of the black dots implies that the black dots were derived from the agglomeration of the initial particles of the Al_2O_3 powder. Such a larger defect leads low mechanical strength, as given by the Griffith Criteria. Even PECS with high pressure, the powder properties such as the agglomeration is very important. PECS under 100 MPa in uniaxial pressure cannot eliminate the agglomeration of the initial particles. In particular elimination of the agglomeration of the initial particles is very important in even PECS for structural ceramics and transparent ceramics.

Density	99.8%
Average Grain Size	0.31 μm
Vickers Hardness	20.8 GPa
Bending Strength	400 MPa
Fracture Toughness	3.3 $MPam^{1/2}$

Table 2. Mechanical Properties of Transparent Al_2O_3 prepared by TS-PECS (1000°C for 60 min, 1200°C for 20 min, 100MPa 100 K/min)

Figure 6 An optical microscopic image of the inside of the transparent Al_2O_3 prepared by TS-PECS

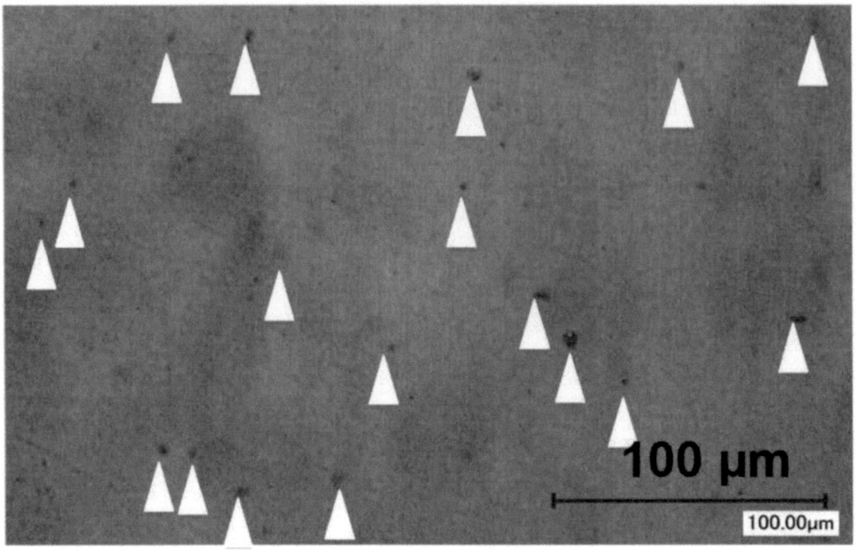

Figure 6. An optical microscopic image of the inside of the transparent Al_2O_3 prepared by TS-PECS

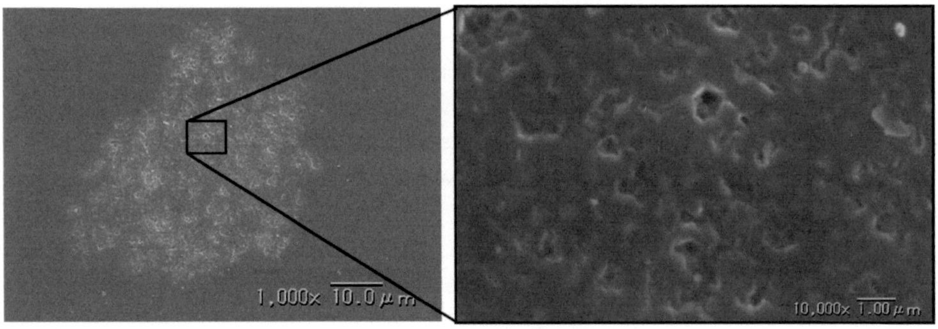

Figure 7. A scanning electron microscopic image of the cross-section of a black dot in transparent Al_2O_3.

TS-PECS is also available for preparing transparent colored-Al_2O_3 and $MgAl_2O_4$. Figure 7 shows appearance of various transparent Al_2O_3 and $MgAl_2O_4$ added with different dopants. A red color in Al_2O_3 and $MgAl_2O_4$ is caused by doping Cr_2O_3. In polycrystalline Al_2O_3, MnO causes the colour of orange or brown, however less transparency. Doping MnO into $MgAl_2O_4$ shows yellow in color and good transparency.

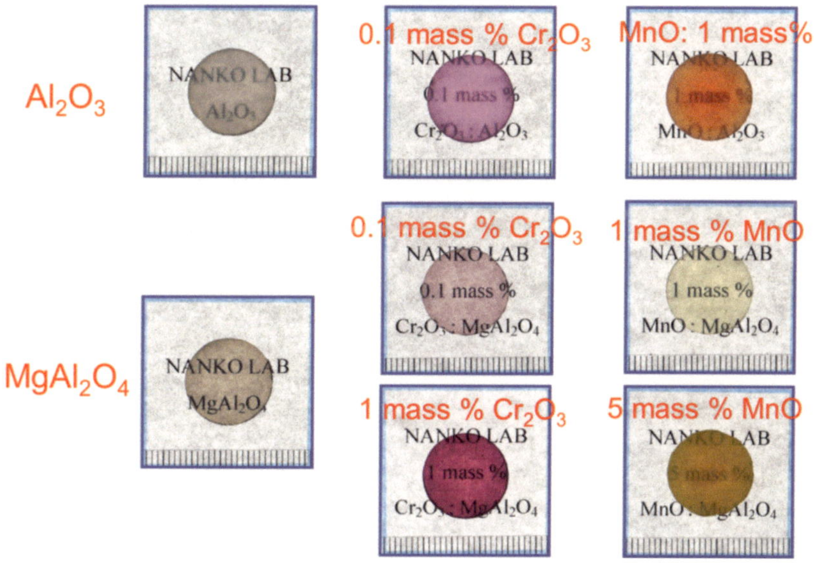

Figure 8. Appearance of various transparent Al_2O_3 and $MgAl_2O_4$ added with different dopants.

6. Summary

Transparent Polycrystalline Al_2O_3 can be produced by using PECS with advanced Al_2O_3 powder. The preferred techniques of PECS to obtain better transparency in Al_2O_3 are classified into the following: (1) high-pressure PECS, (2) slow-heating PECS, (3) fast-heating PECS, and (4) two-step PECS, as well as PECS with preferred additives. Influences in sintering parameters in PECS for transparent polycrystalline Al_2O_3 are still not clear. At least, PECS with slow heating rate and two-step heating profile is preferred to produce transparent polycrystalline Al_2O_3. Using advanced Al_2O_3 powder and PECS, agglomeration of the particles is still significant issue in transparent polycrystalline Al_2O_3. Management of Al_2O_3 powder to reduce the agglomeration is necessary to increase transparency of Al_2O_3 prepare by using PECS.

Acknowledgements

The authors thank to Mr. Masakazu Kawahara (Fuji Electric Industry Co. Ltd.) and Mr. Shinichi Takei (SinterLand Inc.) for useful discussion and Mr. Huu Hien Nguyen who is a graduate student of Nagaoka University of Technology, for conducting experiments on colored Al_2O_3 and $MgAl_2O_4$ and observation of black dots in transparent polycrystalline Al_2O_3.

Author details

Makoto Nanko[1*] and Khanh Quoc Dang[2]

*Address all correspondence to: nanko@mech.nagaokaut.ac.jp

1 Department of Mechanical Engineering, Nagaoka University of Technology, Japan

2 School of Materials Science and Engineering, Hanoi University of Science and Technology, Vietnam

References

[1] Shevell SK. The science of color. Elsevier, Oxford, UK; 2003.

[2] Grasso S., Sakka Y., Maizza G. Electric Current Active/assisted Sintering (ECAS): a Review of Patents 1906-2008. Science and Technology of Advanced Materials 2009; 10 053001/1-24.

[3] Munir ZA., Anselmi-Tamburini U., Oyanagi M. The Effect of Electric Field and Pressure on the Synthesis and Consolidation of Materials: A Review of the Spark Plasma Sintering Method. Journal of Materials Science 2006; 41 763-777.

[4] Duval D'Adrian AL. Article of Fused Metallic Oxide and Process of Producing the Same. US Patent No. 1,430,724 1922.

[5] Bloxam AG. Improved Manufacture of Electric Incandescence Lamp Filaments from Tungsten or Molybdenum or an Alloy Thereof. GB Patent No. 190527002 1906.

[6] Bloxam AG. Improved Manufacture of Filaments of Tungsten or Molybdenum for Electric Incandescence Lamps. GB Patent No. 190609020 1906.

[7] Taylor GF. Improvements in and Relating to Methods of and Apparatus for Producing Hard Metal Compositions. GB Patent No. 385629 1932.

[8] Taylor GF. Apparatus for Making Hard Metal Compositions. US Patent No. 1896854 1933.

[9] Taylor GF. Welding Process. US Patent No. 1896853 1933.

[10] Kim DK., Pak HR., Okazaki K. Electrodischarge Compaction of Nickel Powders. Materials Science and Engineering A 1988; 104 191-200.

[11] Inoue K. Electric-Discharge Sintering. US Patent No. 3241956 1966.

[12] Inoue K. Apparatus for Electrically Sintering Discrete Bodies. US Patent No. 3250892 1966.

[13] Yoo SH., Sethuram KM., Sudarshan TS. Apparatus for Bonding a Particle Material to Near Theoretical Density. US Patent No. 5989487 1999

[14] Tokita M. Trends in Advanced SPS Spark Plasma Sintering Systems and Technology. Journal of the Society of Powder Technology, Japan 1993; 30 790-804.

[15] Inoue K. Method and Apparatus for Electric Discharge Sintering. JP Patent No. 03056604 1991.

[16] Orrù R., Richeri R., Locci AM., Cincotti A., Cao G. Consolidation/synthesis of Materials by Electric Current Activated/assisted Sintering. Materials Science and Engineering R 2009; 63 127-287.

[17] Mirva E. Spark plasma sintering and deformation bihaviour of Titanium and Titanium/TiB2 composites. Licentiate thesis. Stockholm University; 2007.

[18] Matsugi K., Kuramoto H., Hatayama T., Yanagisawa O. Temperature distribution at steady state under constant current discharge in spark sintering process of Ti and Al_2O_3 powders. Journal of Materials Processing Technology 2003; 134 225-232.

[19] Wang YC., Fu ZY., Wang WM., Zhu HX. Temperature Field Distribution in Spark Plasma Sintering of BN. ournal of Wuhan University of Technology 2002; 17 19-21.

[20] Nanko M., Maruyama T., Tomino H. Neck Growth on Initial Stage of Pulse Current Pressure Sintering for Coarse Atomized Powder Made of Cast-iron. Journal of the Japan Institute of Metals 1999; 63 917-923.

[21] Nanko M., Oyaidu T., Maruyama T. Densification of Ni-20Cr alloy course-powder by pulse current pressure sintering. Journal of the Japan Institute of Metals 2002; 66 87-93.

[22] Kim HT., Kawahara M., Tokita M. Specimen temperature and sinterability of Ni powder by spark plasma sintering. Journal of the Japan Society of Powder and Powder Metallurgy 2000; 47(8) 887-891

[23] Wang Y., Fu Z., Zhang Q. SPS Temperature Distribution of Different Conductivity Materials. Key Engineering Materials 2002; 224-226 717-720

[24] Tomino H., Watanabe H., Kondo Y. Electric Current Path and Temperature Distribution for Spark Sintering. Journal of the Japan Society of Powder and Powder Metallurgy 1997; 44 974-979.

[25] Xie G., Ohashi O., Chiba K., Yamaguchi N., Song M., Furuya K., Noda T. Frequency Effect on Pulse Electric Current Sintering Process of Pure Aluminum Powder. Materials Science and Engineering A 2003; 359 384-390.

[26] Chen W., Anselmi-Tamburini U., Garay JE., Groza JR., Munir ZA. Fundamental Investigation on the Spark Plasma Sintering/ Synthesis Process I. Effect of DC Pulsing on Reactivity. Materials Science and Engineering A 2005; 394 132-138.

[27] Anselmi-Tamburini U., Garay JE., Munir ZA. Fundamental Investigations on the Spark Plasma Sintering/ Synthesis Process III. Current Effect on Reactivity", Materials Science and Engineering A 2005; 407 24-30.

[28] Kondo T., Yasuhara M., Kuramoto T., Kodera Y., Ohyanagi M., Munir ZA. Effect of Pulsed DC Current on Atomic Diffusion of Nb-C Diffusion Couple. Journal of Materials Science 2008; 43 6400-6405.

[29] Kondo T., Kuramoto T., Kodera Y., Ohyanagi M., Munir ZA. Influence of Pulsed DC Current and Electric Field on Growth of Carbide Ceramics during Spark Plasma Sintering. Journal of the Ceramic Society of Japan 2008; 116 1187-1192.

[30] Kondo T., Kuramoto T., Kodera Y., Ohyanagi M., Munir ZA. Enhanced Growth of Mo_2C Formed in Mo-C Diffusion Couple by Pulse DC Current. Journal of the Japan Society of Powder and Powder Metallurgy2008; 55 643-650.

[31] Inoue K. Electric Discharge Heat Treatment of Metals in Electrolytes. US Patent No. 3,188,245; 1965.

[32] Nishimoto K., Saida KJ., Tsuzuki R. Effect of Pulsed Electric-current on Densification Behavior of Bonded Interlayer of Oxide-dispersion-strengthened Super Alloys Joint. Journal of the Japan Institute of Metals 2001; 65(8) 747-755.

[33] Dang QK., Nanko M., Kawahara M., Takei S. Densification of Alumina Powder by Using PECS Process with Different Pulse Electric Current Waveforms. Materials Science Forum 2009; 620-622 101-104.

[34] Dang QK., Nanko N. Effects of ON/OFF Pulse pattern on Sintering Alumina by a Pulsed Electric Current Sintering Process. Journal of Ceramic Processing Research 2009; 10 s32-s38.

[35] Dang QK., Kawahara M., Takei S., Nanko M. Effects of Pulsed Current Waveforms on Sample Temperature and Sintering Behavior in PECS of Alumina. Journal of the Japan Society of Powder and Powder Metallurgy 2009; 56 780-787.

[36] Anselmi-Tamburini U., Gennari S., Garay JE., Munir ZA. Fundamental Investigations on the Spark Plasma Sintering/ Synthesis Process II. Modeling of Current and Temperature Distributions. Materials Science and Engineering A 2005; 394 139-148.

[37] Zhang DM., Fu ZY., Wang YC., Zhang QJ., Guo JK. Heterogeneous of Non-conductive Materials Sintering by Pulse Electric Current. Key Engineering Materials 2002; 224-226 729-734.

[38] Peelen JGJ., Metselaar R. Light Scattering by Pores in Polycrystalline Materials: Transmission Properties of Alumina. Journal of Applied Physics 1974; 45 216-220.

[39] Apetz R., Bruggen MPB. Transparent Alumina: a Light-scattering Model. Journal of the American Ceramic Society 2003; 86 480-486.

[40] Pecharromàn C., Mata-Osoro G., Antonio Dìaz L., Torrecillas R., Moya JS. On the transparency of nanostructured alumina: Rayleigh-Gans Model for Anisotropic Spheres. Optics Express 2009; 17 6899-6912.

[41] Mizuta H., Oda K., Shibasaki Y., Maeda M., Machida M., Ohshima K. Preparation of High-Strength and Translucent Alumina by Hot Iso-Static Pressing. Journal of the American Ceramic Society 1992; 75 469-73.

[42] Coble RL. Transparent Alumina and Method of Preparation. U.S. Patent No. 3,026,210 1962.

[43] Wei GC., Rhodes WH. Sintering of Translucent Alumina in a Nitrogen–Hydrogen Gas Atmosphere. Journal of the American Ceramic Society 2000; 83 1641-8.

[44] Mao XJ., Wang SW., Shimai S., Guo JK. Transparent Polycrystalline Alumina Ceramics with Orientated Optical Axes. Journal of the American Ceramic Society 2008; 91 3431-3.

[45] Krell A., Blank P., H. Ma, Hutzler T. Transparent Sintered Corundum with High Hardness and Strength. Journal of the American Ceramic Society 2003; 86 12-18.

[46] Bernard-Granger G., Guizard C. Influence of MgO or TiO2 Doping on the Sintering Path and on the Optical Properties of a Submicronic Alumina Material. Scripta Materialia 2007; 56 983-986.

[47] Kim DS., Lee JH., Sung RJ., Kim SW., Kim HS., Park JS. Improvement of Translucency in Al_2O_3 Ceramics by Two-step Sintering Technique. Journal of the European Ceramic Society 2007; 27 3629-3632.

[48] Peelen JGJ., Metselaar R. Light Scattering by Pores in Polycrystalline Materials: Transmission Properties of Alumina. Journal of Applied Physics 1974; 45 216-220.

[49] Peelen JGJ. Light Transmission of Sintered Alumina. Philips Technical Review 1976; 36 47-52.

[50] Peelen JGJ. Transparent Hot-pressed Alumina. I: Hot Pressing of Alumina. Ceramurgia International 1979; 5 70-75.

[51] Peelen JGJ. Transparent Hot-pressed Alumina. II: Transparent versus Translucent Alumina. Ceramurgia International 1979; 5 115-119.

[52] Krell A., Klimke J., Hutzler T. Advanced Spinel and Sub-μm Al_2O_3 for Transparent Armour Applications. Journal of the European Ceramic Society 2009; 29 275-281.

[53] Jiang D., Hulbert DM., Anselmi-Tamburini U., Ng T., Land D., Mukherjee AK. Optically Transparent Polycrystalline Al_2O_3 Produced by Spark Plasma Sintering. Journal of the American Ceramic Society 2008; 91 151-154.

[54] Aman Y., Garnier V., Djurado E. Influence of Green State Processes on the Sintering Behaviour and the Subsequent Optical Properties of Spark Plasma Sintered Alumina. Journal of the European Ceramic Society 2009; 29 3363-3370.

[55] Suárez M., Fernández A., Menéndez J.L., Torrecillas R. Grain Growth Control and Transparency in Spark Plasma Sintered Self-doped Alumina Materials. Scripta Materialia 2009; 61 931-934.

[56] Stuer M., Zhao Z., Aschauer U., Bowen P. Transparent Polycrystalline Alumina Using Spark Plasma Sintering: Effect of Mg, Y and La doping. Journal of the European Ceramic Society 2010; 30 1335-1343.

[57] Kim BN., Hiraga K., Morita K., Yoshida H. Spark Plasma Sintering of Transparent Alumina. Scripta Materialia 2007; 57 607-610.

[58] Kim BN., Hiraga K., Morita K., Yoshida H. Effects of Heating Rate on Microstructure and Transparency of Spark Plasma Sintered Alumina. Journal of the European Ceramic Society 2009; 29 323-327.

[59] Kim BN., Hiraga K., Morita K., Yoshida H., Miyazaki T., Kagawa Y. Microstructure and Optical Properties of Transparent Alumina. Acta Materialia 2009; 57 1319-1326.

[60] O YT., Koo J., Hong KJ., Park JS., Shin DC. Effect of Grain Size on Transmittance and Mechanical Strength of Sintered Alumina. Materials Science and Engineering A 2004; 374 191-195.

[61] Pecharromàn C., Mata-Osoro G., Antonio Dìaz L., Torrecillas R., Moya JS. On the transparency of nanostructured alumina: Rayleigh-Gans Model for Anisotropic Spheres. Optics Express 2009; 17 6899-6912.

[62] Braun A., Falk G., Clasen R. Transparent Polycrystalline Alumina Ceramic with Submicrometre Microstructure by Means of Electrophoretic Deposition. Material Wissenschaft und Werkstofftechnik 2006; 37 293-297.

[63] Krell A., Blank P., Ma H., Hutzler T. and Nebelung M. Processing of High-Density Submicrometer Al_2O_3 for New Applications. Journal of the American Ceramic Society 2003; 86 546-553.

[64] Krell A., Klimke J. Effects of the Homogeneity of Particle Coordination on Solid-state Sintering of Transparent Alumina. Journal of the American Ceramic Society 2006; 89 1985-1992.

[65] Krell A., Blank P. Grain size Dependence of Hardness in Dense Submicrometer Alumina. Journal of the American Ceramic Society 1995; 78 (4) 1118-1120.

[66] Cheng J., Agarwal D., Zhang Y., Roy R. Microwave Sintering of Transparent Alumina. Materials Letters 2002; 56(4) 587-592.

[67] Kwon OH., Nordahl CS., Messing GL. Submicrometer Transparent Alumina by Sinterforging Seeded γ-alumina Powders. Journal of the American Ceramic Society 1995; 78 491-494.

[68] Grasso S., Kim BN., Hu C., Maizza G., Sakka Y. Highly Transparent Pure Alumina Fabricated by High-pressure Spark Plasma Sintering. Journal of the American Ceramic Society 2010; 93 2460-2462.

[69] Hayashi K., Kobayashi O., Toyoda S., Morinaga K. Transmission Optical Properties of Polycrystalline Alumina with Submicron Grains. Materials Transactions, JIM 1991; 32 1024-1029.

[70] Aman Y., Garnier V., Djurado E. A Screening Design Approach for the Understanding of Spark Plasma Sintering Parameter: A Case of Translucent polycrystalline Undoped Alumina. International Journal of Applied Ceramic Technology 2010; 7 574-586.

[71] Chakravarty D., Bysakh S., Muraleedharan K., Rao T.N., Sundaresan R. Spark Plasma Sintering of Magnesia-doped Alumina with High Hardness and Fracture Toughness. Journal of the American Ceramic Society 2008; 91 203-208.

[72] Park CW., Yoon DY. Effects of SiO2, CaO2, and MgO additions on the Grain Growth of Alumina. Journal of the American Ceramic Society 2000; 83 2605-609.

[73] Dobedoe RS., West GD., Lewis MH. Spark Plasma Sintering of Ceramics. Bulletin of the European Ceramic Society 2003; 1 19-24.

[74] Makino Y, Kawahara M, Sakaguchi M, Kaiyama Y, Akatsuka K, Yasuno T, J. Jpn. Powder Powder Metall., 2012; 59 (9) 532-537.

[75] Dang K. Q, Nanko M, transparent alumina prepared by two-step sintering in pulsed electric current sintering. in PM2012: proceedings of PM2012, 14-18 October 2012, Yokohama, Japan; 2013.

[76] Nanko M, Dang K. Q., Two-step Pulsed Electric Current Sintering of Transparent Al_2O_3 Ceramics, Adv. Appl. Ceram.2014; 113 (2) 80-84.

[77] Wang C, Zhao Z, Transparent Polycrystalline Ruby Ceramic by Spark Plasma Sintering, Mater. Res. Bull. 2010; 45 (9) 1127-1131.

[78] Dang K. Q., Kawahara M, Takei S, Nanko M, Fabrication of Transparent Cr-Doped Al2O3 Made by Pulsed Electric Current Sintering Process, Ceram. Intl. 2011; 37 957-963.

[79] Roussel N, Lallemant L, Durand B, Guillemet S, Chane Ching J.-Y., Fantozzi G, Garnier V, Bonnefont G, Effects of the Nature of the Doping Salt and of the Thermal Pretreatment and Sintering Temperature on Spark Plasma Sintering of Transparent Alumina, Ceram. Intl. 2011; 37 (8) 3565-3573.

[80] Kim B.N., Hiraga K, Grasso S, Morita K, Yoshida H, Zhang H, Sakka Y, High-pressure Spark Plasma Sintering of MgO-doped Transparent Alumina, J. Ceram. Soc. Jpn. 2012; 120 (3) 116-118.

[81] Lallemant L,Fantozzi G, Garnier V, Bonnefont G, Transparent Polycrystalline Alumina Obtained by SPS: Green Bodies Processing Effect, J. Euro. Ceram. Soc. 2012; 32 2909-2915.

[82] Roussel N, Lallemant L, Chane-Ching J.-Y., Guillemet-Fristch S, Durand B, Garnier V, Bonnefont G, Fantozzi G, Bonneau L, Trombert S, Garcia-Gutierrez D, Highly Dense, Transparent α-Al_2O_3 Ceramics From Ultrafine Nanoparticles Via a Standard SPS Sintering, J. Amer. Ceram. Soc., 2013; 96 (4) 1039-1042.

[83] Anselmi-Tamburini U, Garay J.E., Munir Z.A., Fast Low-temperature Consolidation of Bulk Nanometric Ceramic Materials, Scripta Mater. 2006; 54 823-828.

[84] Kim B.N., Hiraga K, Grasso S, Morita K, Yoshida H, Zhang H, Sakka Y, High-pressure Spark Plasma Sintering of MgO-doped Transparent Alumina, J. Ceram. Soc. Jpn. 2012; 120 (3) 116-118.

[85] Zhang H, Kim B.-N. Morita K, Yoshida H, Hiraga K, Sakka Y, Fabrication of Transparent Yttria by High-Pressure Spark Plasma Sintering, J. Amer. Ceram. Soc., 2011; 94 (10) 3206–3210.

[86] Morita K, Kim B.N., Hiraga K, Yoshida H, Fabrication of transparent $MgAl_2O_4$ Spinel polycrystal by spark plasma sintering processing, Scripta Mater.2008; 58, 1114-1117.

[87] Kim B.N., Hiraga K, Morita K, Yoshida H, Park Y. J., Sakka Y, Dynamic Grain Growth during Low-temperatrue Spark Plasma Sintering of Alumina, Scripta Mater. 2014; 80() 29-32.

[88] An L, Ito A, Goto T, Two-step Pressure Sintering of Transparent Lutetium Oxide by Spark Plasma Sintering, J. Euro Ceram. Soc.2011; 31 1597–1602.

[89] Gasso S, Kim B. N., Hu C, Mizza G, Sakka Y, Highly Transparent Pure Alumina Fabricated by High-Pressure Spark Plasma Sintering. J. Amer. Ceram. Soc. 2010; 93 (9) 2460-24 62.

[90] Frage N, Cohen S, Meir S, Kalabukhov S, Dariel M. P., Spark Plasma Sintering (SPS) of Transparent Magnesium-aluminate Spinel, J. Mater. Sci. 2007; 42 3273-3275.

[91] Bernard-Granger G, Benameur N, Guizard C, Nygren M, Influence of Graphite Contamination on the Optical Properties of Transparent Spinel Obtained by Spark Plasma Sintering, Script Mater. 1009; 60 164-167.

[92] Wang C, Zhao Z, Transparent $MgAl_2O_4$ Ceramic Produced by Spark Plasma Sintering, Scripta Mater.2009; 61 (2), 193-196.

[93] Bonnefonta G, Fantozzi G, Trombert S, Bonneau L, Fine-grained Transparent $MgAl_2O_4$ Spinel Obtained by Spark Plasma Sintering of Commercially Available Nanopowders, Ceram. Intl. 2012; 38 (1) 131-140.

[94] Fu P, Lu W, Lei W, Xu Y, Wang X, Wu J, Transparent Polycrystalline $MgAl_2O_4$ Ceramic Fabricated by Spark Plasma Sintering: Microwave Dielectric and Optical Properties, Ceram. Intl. 2013: 39 (3) 2481-2487.

[95] Casolco S. R., Xu J, Garay J. E., Transparent/translucent Polycrystalline Nanostructured Yttria Stabilized Zirconia with Varying Colors, Scripta Mater. 2008; 58, 516-519.

[96] Alaniz J. E., Perez-Gutierrez F. G., Aguilar G, Garay J. E., Optical Properties of Transparent Nanocrystalline Yttria Stabilized Zirconia, Opt. Mater. 2009; 32 62–68.

[97] Lei L.W., Fu Z.Y., Wang H, Lee S.W., Niihara K., Transparent Yttria Stabilized Zirconia from Glycine-nitrate Process by Spark Plasma Sintering, Ceram. Intl., 2012; 38 (1) 23-28.

[98] Zhang H, Kim B.-N. Morita K, Yoshida H, Hiraga K, Sakka Y, Fabrication of Transparent Yttria by High-Pressure Spark Plasma Sintering, J. Amer. Ceram. Soc., 2011; 94 (10) 3206–3210.

[99] An L, Ito A, Goto T, Transparent Yttria Produced by Spark Plasma Sintering at Moderate Temperature and Pressure Profiles, J. Euro. Ceram. Soc. 2012; 32 (5) 1035–1040.

[100] An L, Ito A, Goto T, Fabrication of Transparent Lutetium Oxide by Spark Plasma Sintering, J. Amer. Ceram. Soc. 2011; 94 (3) 695-698.

[101] Prakasam M, Viraphong O, Michau D, Veber P, Velázquez M, Shimamura K, Largeteau A, Yb^{3+} Doped Lu_2O_3 Transparent Ceramics by Spark Plasma Sintering, Ceram. Intl. 2013; 39 1307-1313.

[102] Chaim R, Shen Z, Nygren M, Transparent Nanocrystalline MgO by Rapid and Low-temperature Spark Plasma Sintering, J. Mater. Res. 2004; 19 (4) 2527-2531.

[103] Tran T. B., Hayun S, Navrotsky A, Castro R. H. R., Transparent Nanocrystalline Pure and Ca-Doped MgO by Spark Plasma Sintering of Anhydrous Nanoparticles, J. Amer. Ceram. Soc. 2012; 95 (4) 1185-1188.

[104] An L, Ito A, Goto T, Highly Transparent Lutetium Titanium Oxide Produced by Spark Plasma Sintering, J. Euro. Ceram. Soc. 2011; 31 237-240.

[105] An L, Ito A, Goto T, Fabrication of Transparent Lu_3NbO_7 by Spark Plasma Sintering, Mater. Lett. 2011; 65 3167-3169.

[106] An L, Ito A, Goto T, Fabrication of Transparent $La_2Zr_2O_7$ by Reactive Spark Plasma Sintering, Key Eng. Mater. 2011; 484 135-138.

[107] Kawagoe D, Ioku K, Fujimori H, Goto S, Transparent β-Tricalcium Phosphate Ceramics Prepared by Spark Plasma Sintering, J. Ceram. Soc. Jpn. 2004; 112 (8) 462-463.

[108] Chaim R, Marder-Jaeckel R., Shen J. Z., Transparent YAG Ceramics by Surface Softening of Nanoparticles in Spark Plasma Sintering, Mater. Sci. Eng. A 2006; 429 (1/2) 74-78.

[109] Frage N, Kalabukhov S, Sverdlov N, Ezersky V, Dariel M. P., Densification of Transparent Yttrium Aluminum Garnet (YAG) by SPS Processing, J. Euro. Ceram. Soc. 2010; 30 3331-3337.

[110] Spina G, Bonnefont G, Palmero, Fantozzi G, Chevalier J, Montanaro L, Transparent YAG Obtained by Spark Plasma Sintering of Co-precipitated Powder. Influence of Dispersion Route and Sintering Parameters on Optical and Microstructural Characteristics, J. Euro. Ceram. Soc. 2012; 32 2957-2964.

[111] Chesnaud A, Bogicevic C, Karolak F, Estournes C, Dezanneau G, Preparation of Transparent Oxyapatite Ceramics by Combined Used of Freeze-drying and Spark-plasma Sintering, Chem. Commun. 2007 1550-1552.

[112] Eriksson M, Liu Y, Hu J, Gao L, Nygren M, Shen Z, Transparent hydroxyapatite Ceramics with Nanograin Structure Prepared by High Pressure Spark Plasma Sintering at the minimized Sintering Temperature, J. Euro. Ceram. Soc. 2011; 31 1533-1540.

[113] Zhang G, Wang Y, Fu Z, Wang H, Wang W, Zhang J, Lee S.-W., Niihara K, Transparent Mullite Ceramic from Single-Phase Gel by Spark Plasma Sintering, J. Euro. Ceram. Soc. 2009; 29 2705-2711.

Chapter 2

Spark Plasma Sintering of Negative Temperature Coefficient Thermistor Ceramics

Aimin Chang, Bo Zhang, Yiquan Wu, Qing Zhao,
Huimin Zhang, Jincheng Yao, Jinbao Xu and
Pengjun Zhao

Additional information is available at the end of the chapter

http://dx.doi.org/10.5772/58496

1. Introduction

Spark plasma sintering (SPS), also known as field assisted sintering technique (FAST), belongs to a class of sintering techniques in which densification is enhanced by the simultaneous application of axial pressure and elevated temperature generated by a high current flow [1, 2]. SPS has been successfully used in the preparation of functionally graded materials [3], ceramics [4-6], magnetic materials [7, 8], alloy [9], etc. It is believed that the SPS has a significant advantages in decreasing the sintering temperature and sintering periods, and also preparing high density ceramics [10]. Although SPS has been widely used to prepare dense transparent and structural ceramics, little reports focus on the application of this technique to prepare dense ceramics for negative temperature coefficient (NTC) thermistor applications [11,12]. So this chapter first provides a summary of fundamental theoretical aspects of spark plasma sintering, and then the application of spark plasma sintering in NTC thermistor ceramics are introduced emphatically. At the end, the future research and application of spark plasma sintering in the NTC thermistor ceramics are forecasted.

2. Theoretical aspects of spark plasma sintering

SPS is a novel effective sintering technique, which has been used in the research and development of various kinds of materials. However, there is no uniform understanding in the sintering mechanism of SPS. In general, SPS is a pressure sintering method based on the simultaneous application of axial pressure and high temperature plasma momentarily generated in the gaps between powder materials by electrical discharge at the beginning of ON-OFF DC pulse energizing [13]. In addition to have the Joule heating due to the electric current and plastic deformation produced by pressure, SPS also generates DC pulse voltage between the powder particles, and effectively makes use of the spontaneous heat generated by the discharge between powder particles, thus resulting in some special characteristics. Compared to the conventional sintering, SPS has two important characteristics [14, 15]: (1) SPS process can make high-energy pulse focus on the grain junction point, thus saving the energy; (2) A high energy, low voltage spark pulse current momentarily generates spark plasma and produces a high localized temperature from several to ten thousand °C between the particles and then resulting in optimum thermal diffusion and grain boundary migration, i.e. more material transfer can be intensified and thus high density ceramics can be obtained through spark plasma sintering with a low sintering temperature and a short sintering period. Fig. 1 shows the schematic of SPS furnace [16].

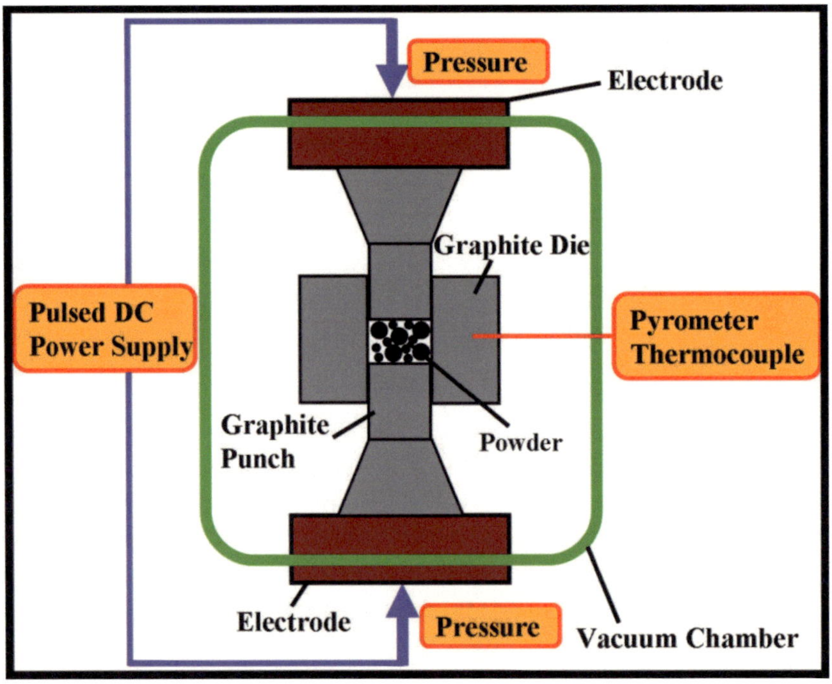

Figure 1. Schematic of SPS furnace [16].

3. Application of spark plasma sintering in NTC thermistor ceramics

In the past decades, more advanced techniques such as microwave sintering [17, 18], and nitrogen atmosphere sintering [19] have been used for NTC ceramic powder consolidation. However, there are few reports focusing on the application of SPS technique to prepare dense ceramics for NTC thermistor applications. The advantages of spark plasma sintering against conventional sintering for high temperature NTC thermistor ceramics are reviewed as follows.

3.1. Brief introduction of NTC thermistors

NTC thermistors are thermally sensitive resistors whose resistance decreases with increasing temperature. Their resistivity can be expressed by the following Arrhenius equation [20, 21]: $\rho=\rho_o\exp(Ea/kT)$, where ρ_o is the resistivity of the material at infinite temperature, T is the absolute temperature, Ea is the activation energy for electrical conduction, and k is the Boltzmann constant. They are mainly used in industrial areas as elements for temperature measurements, control, etc [22].

In the past few years, development of novel high temperature NTC thermistor materials has been motivated by the requirements of particle filters and catalytic converters in exhaust pipe for automotive motors [23, 24]. NTC thermistor ceramics composed of spinel structure (MMn_2O_4, where M=Ni, Co, Fe, Cu, Zn) show aging of the electrical properties and their application is commonly limited to temperatures below 300°C [25, 26]. The literature suggests that rare earth (Sm, Tb, Y, etc) perovskite oxides (ABO_3) can be used for measurements from ambient to 1000°C [23]. In particular, $YCrO_3$ having an orthorhombic perovskite structure, has been considered as a candidate for high temperature NTC thermistor applications [19, 23, 27, 28]. However, the material shows poor sinterability and is difficult to densify under ambient atmospheric conditions or through pressureless sintering techniques [29, 30]. We have investigated the spark plasma sintering of $YCr_{1-x}Mn_xO_3$ ceramics and $MgAl_2O_4$-$YCr_{0.5}Mn_{0.5}O_3$ composite ceramics, and their NTC electrical properties.

3.2. Spark plasma sintering and electrical properties of $YCr_{1-x}Mn_xO_3$ NTC ceramics

In the conventional sintering processes, extremely high sintering temperatures (up to 1600°C) and long holding time (several hours) in air are applied in the fabrication of $YCrO_3$ ceramics to achieve the highest density and minimum porosity. The poor sinterability of $YCrO_3$ material is attributed to the loss of Cr_2O_3 through its volatility during sintering process [30, 31]. In our previous work [32, 33], $YCr_{1-x}Mn_xO_3$ ($0\leq x\leq 0.5$) NTC ceramics with a high relative density have been obtained by combing the Pechini method synthesis and SPS. Fig.2 shows the flow chart for the fabrication of $YCr_{1-x}Mn_xO_3$ thermistor powders by a Pechini method. The molar ratio of citric acid, ethylene glycol and metal ions was 1.5:1.5:1. The spark plasma sintering was carried out in vacuum (6 Pa) with an apparatus (FCT Systeme GmbH, FCT, Rauenstein, Germany). The SPS equipment used in the experiments is shown in Fig. 3. Fig.4 shows the time dependence of the temperature and applied pressure during SPS process. The sintering temperature was 1300 ∘C, and the dwell time was 10 min. Fig.5 shows the XRD patterns of the SPS-sintered $YCr_{1-x}Mn_xO_3$ ceramics. All samples had a pure orthorhombic perovskite phase isomorphic to $YCrO_3$ (JCPDF 34-0365) described by the space group *Pnma*, and no secondary phase occurred with the increase of Mn concentration. Fig.6 shows the SEM images of surface section of SPS-sintered $YCr_{1-x}Mn_xO_3$ ceramics. It can be seen that as-sintered $YCrO_3$ ceramics were highly dense, and had a bulk density of 5.6112 g/cm^3, corresponding to 97.6% of the theoretical density (5.751 g/cm^3) [33]. One can observe that grain size decreased with the increase of Mn content. This result may be due to the dragging effect between Mn ions and grain boundary, which increases the energy for the movement of grain boundary and retards the grain growth [34, 35]. The $YCr_{1-x}Mn_xO_3$ NTC thermistor over a wide temperature range of 25 to 300 °C showed a linear relationship between the logarithm of the resistivity and the reciprocal of the absolute temperature. And the resistivity increased at first and then decreased with increasing Mn contents, which had the same varying tendency with activation energy. This electrical conductivity anomaly has been revealed by using defect chemistry theory combination with X-ray photoelectron spectroscopy analysis [32]: The major carriers in $YCrO_3$ are holes. Mn ions are acting as an n-type dope, and partly compensate for the effect of metal vacancies, thus leading to an increase in the resistivity. Mn^{4+} ions increase as Mn content increases from 0.2 to 0.5, which promote the rise in charge carriers and the electron hopping, thereby resulting in a decrease in the resistivity. Therefore, SPS has shown significant advantages against conventional sintering in the fabrication of high density $YCr_{1-x}Mn_xO_3$ ceramics, and provides efficient and viable means for the study of the conduction mechanism of NTC thermistors.

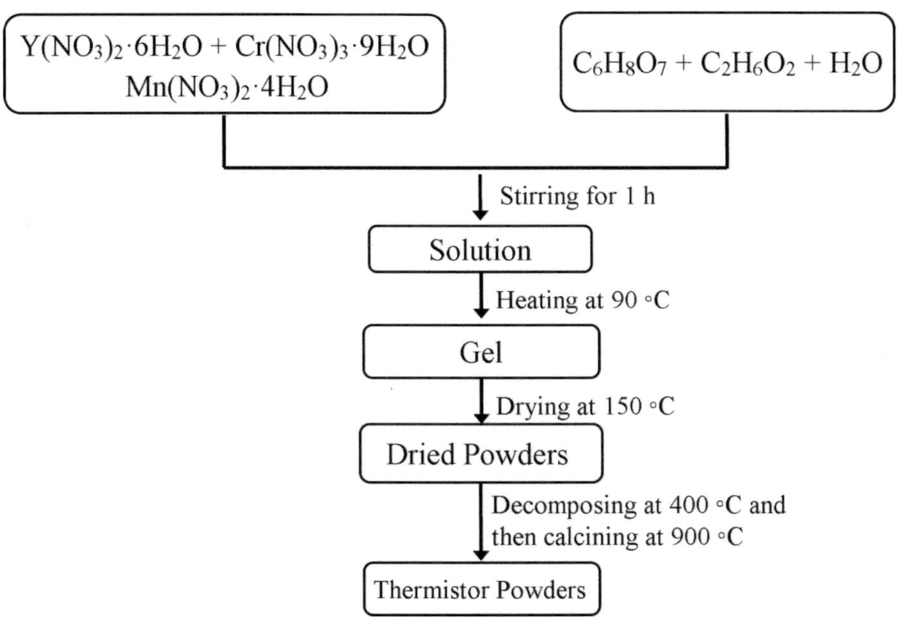

Figure 2. Flow chart for the fabrication of $YCr_{1-x}Mn_xO_3$ thermistor powders by a Pechini method.

http://dx.doi.org/10.5772/58496

Figure 3. Image of SPS equipment (From Alfred University).

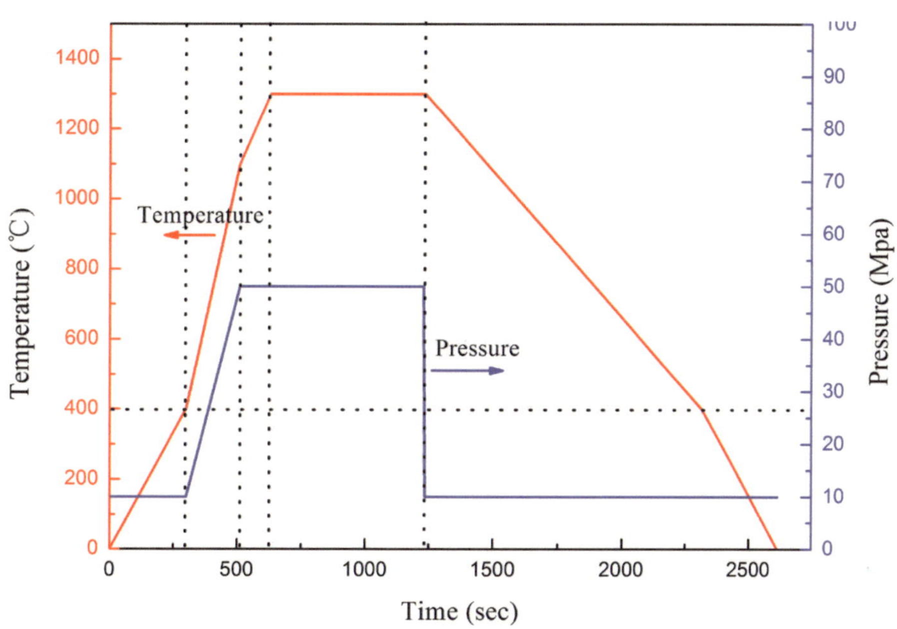

Figure 4. Time dependence of the temperature and applied pressure during SPS process [33].

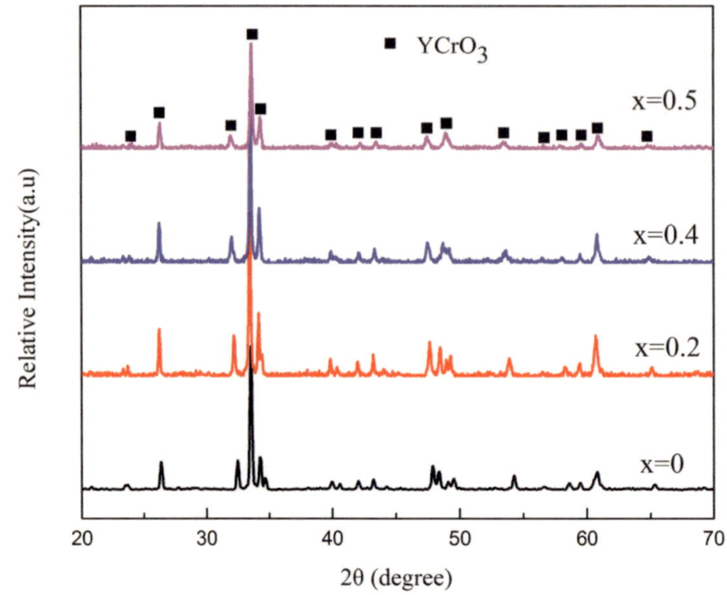

Figure 5. XRD patterns of the SPS-sintered $YCr_{1-x}Mn_xO_3$ ceramics [32].

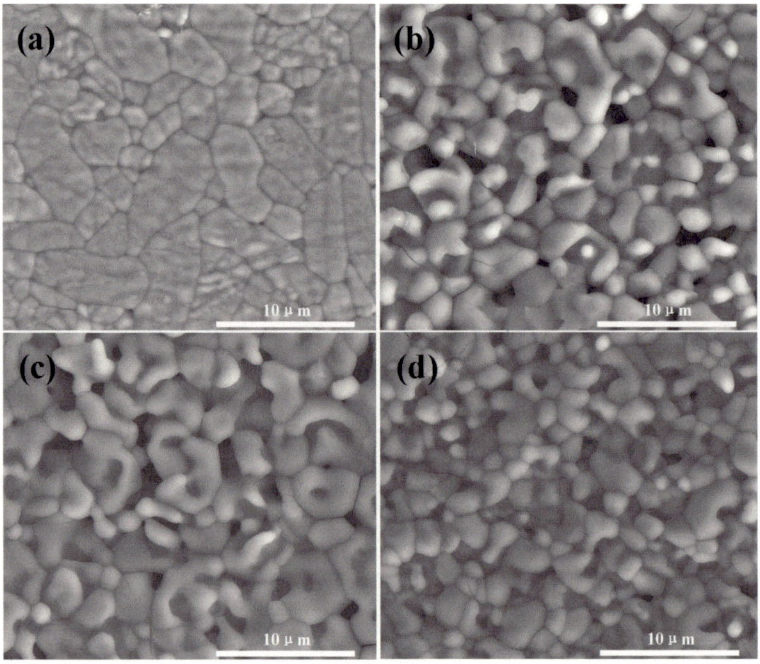

Figure 6. SEM images of surface section of SPS-sintered $YCr_{1-x}Mn_xO_3$ ceramics: (a) x=0; (b) x=0.2; (c) x=0.4; (d) x=0.5.

http://dx.doi.org/10.5772/58496

3.3. Spark plasma sintering and electrical properties of $MgAl_2O_4$-$YCr_{0.5}Mn_{0.5}O_3$ composite NTC ceramics

Recently, there is an increasing interest in exploring NTC behavior in composite materials because of their high temperature potential for combining properties that are difficult to attain separately with the individual component [24]. We have designed and prepared $x$$MgAl_2O_4$-(1-$x$)$YCr_{0.5}Mn_{0.5}O_3$ high temperature composite thermistor ceramics by associating a less resistive phase $YCr_{0.5}Mn_{0.5}O_3$ with a high resistive $MgAl_2O_4$ combination with spark plasma sintering [12]. Fig.7 shows the time dependence of the temperature and applied pressure during SPS process. The sintering temperature was 1200∘C, and the dwell time was 20 min. The SPS-sintered composite ceramics consisted of a cubic spinel $MgAl_2O_4$ phase and an orthorhombic perovskite $YCr_{0.5}Mn_{0.5}O_3$ phase isomorphic to $YCrO_3$. Fig. 8 exhibits the microstructures of the SPS-sintered samples. The SPS-sintered ceramics were highly dense, and their grain sizes were ranging from 0.5 to 2 μm. The relative densities were 95.5%, 97.4% and 94.1% of the theoretical density for x=0.1, 0.4, 0.6, respectively. The resistivity of composite ceramics decreased with increasing temperature from 25 to 1000 ∘C, indicative of NTC characteristics. The obtained ρ_{25}, $B_{25\text{-}150}$, $B_{700\text{-}1000}$, $Ea_{25/150}$ and $Ea_{700/1000}$ of the SPS-sintered composite NTC thermistors were in the range of 1.53×10^6-9.92×10^9 Ωcm, 3380-5172 K, 7239-9543 K, 0.291~0.446 eV, 0.624~0.823 eV, respectively. This result indicates that these values can be adjusted by changing $MgAl_2O_4$ content. Fig. 9 compares the temperature dependence of electrical resistivity ρ of the samples 0.4$MgAl_2O_4$-0.6$YCr_{0.5}Mn_{0.5}O_3$sintered by conventional sintering (CS) and SPS. It can be seen that the samples from SPS-sintered ceramics possessed a higher resistivity than that from conventional sintered ceramics. there are two possible reasons for the increase in the resistivity of SPS-sintered samples [12]: (1) During the SPS process, the short sintering period is advantageous in reducing chromium volatilization, thus leading to a decrease in Cr^{4+} and Mn^{4+} ion concentration, thereby increasing the resistivity as a result; (2) SPS-sintered

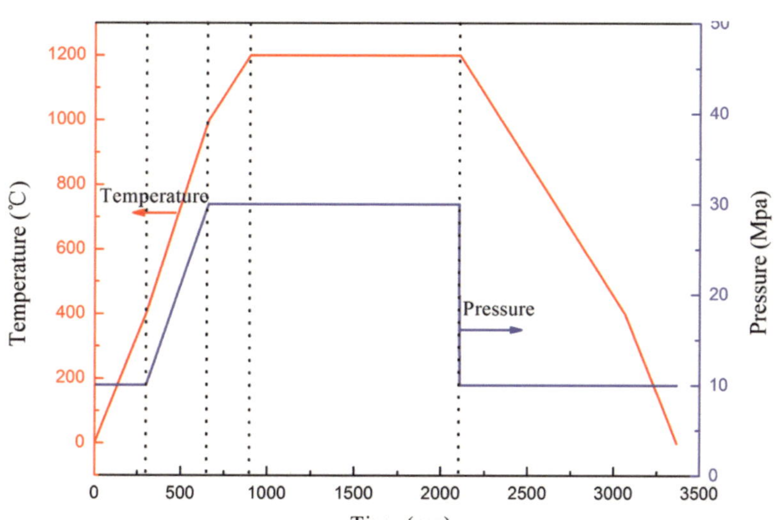

Figure 7. Time dependence of the temperature and applied pressure during SPS process.

samples have a smaller grain size, resulting in a decrease in the time between electron scattering events of charge carriers and thus increasing the resistivity [36]. In conclusion to this, SPS has potential superiority on synthesis high temperature NTC ceramic materials.

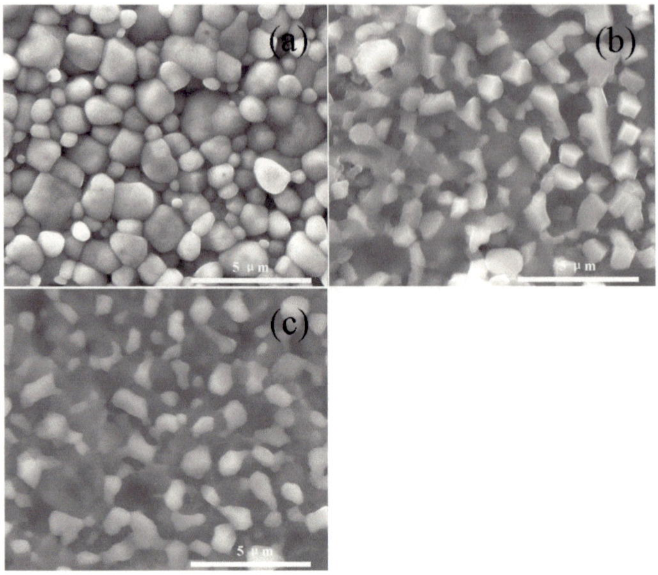

Figure 8. SEM images of the SPS-sintered xMgAl$_2$O$_4$-(1-x)YCr$_{0.5}$Mn$_{0.5}$O$_3$ composite ceramics: (a) x=0.1; (b) x=0.4; (c) x=0.6 [12].

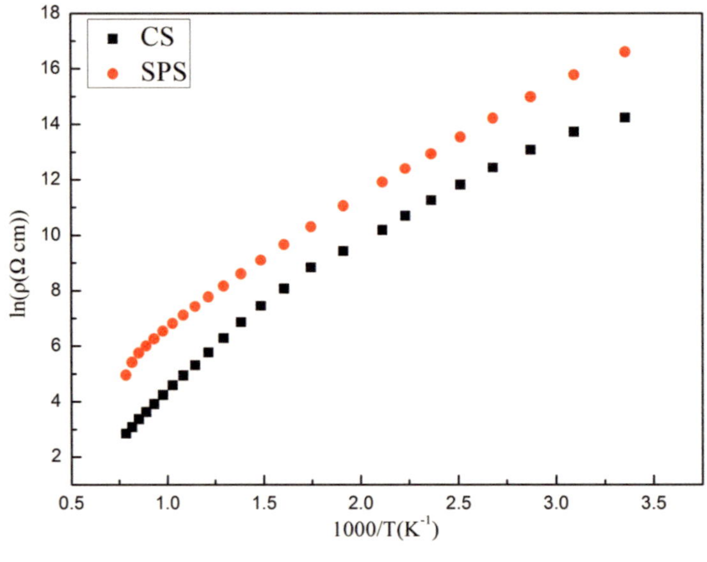

Figure 9. Temperature dependence of electrical resistivity ρ of the samples 0.4MgAl$_2$O$_4$-0.6YCr$_{0.5}$Mn$_{0.5}$O$_3$ sintered by conventional sintering (CS) and SPS [12].

4. Summary

The fundamentals, applications of spark plasma sintering to thermistor ceramics are reviewed in this chapter. For thermistor ceramics, the advantages of spark plasma sintering against traditional sintering are as follows: (1) Faster heating rate decreases the sintering time by using spark plasma sintering, and thus saving energy than traditional sintering; (2) Spark plasma sintering can ensure ceramics with high density and small grain at a low sintering temperature and a short sintering period. What's more, spark plasma sintering has shown significant advantages in the fabrication of $YCrO_3$ perovskite-based thermistor ceramics and provides an efficient mean for the study of the conduction mechanism of NTC thermistors.

In the past years, there have had a significant developments and advances regarding to the spark plasma sintering of ceramics. However, the fundamental theory of spark plasma sintering is not fully understood, and also needs massive fundamental research and practical innovation to perfect. It can be forecasted that there is a great future for the successful commercialization of spark plasma sintering in ceramic preparation.

Author details

Aimin Chang[1*], Bo Zhang[1], Yiquan Wu[3], Qing Zhao[1,2], Huimin Zhang[1], Jincheng Yao[1], Jinbao Xu[1] and Pengjun Zhao[1,2]

*Address all correspondence to: changam@ms.xjb.ac.cn

1 Key Laboratory of Functional Materials and Devices for Special Environments of CAS; Xinjiang Key Laboratory of Electronic Information Materials and Devices; Xinjiang Technical Institute of Physics & Chemistry of CAS, Urumqi, China

2 University of Chinese Academy of Sciences, Beijing, China

3 Kazuo Inamori School of Engineering, New York State College of Ceramics at Alfred University, Alfred, NY, USA

References

[1] S. Meir, S. Kalabukhov, N. Froumin, M.P. Dariel, N. Frage, Synthesis and densification of transparent magnesium aluminate spinel by SPS processing, Journal of the American Ceramic Society, 92 (2009) 358-364.

[2] K. Vanmeensel, A. Laptev, O. Van der Biest, J. Vleugels, Field assisted sintering of electro-conductive ZrO_2-based composites, Journal of the European Ceramic Society, 27 (2007) 979-985.

[3] R. Watanabe, T. Nishida, T. Hirai, Present status of research on design and processing of functionally graded materials, Metals and Materials International, 9 (2003) 513-519.

[4] N. Gao, J. Li, D. Zhang, Y. Miyamoto, Rapid synthesis of dense Ti_3SiC_2 by spark plasma sintering, Journal of the European Ceramic Society, 22 (2002) 2365-2370.

[5] J.G. Noudem, D. Kenfaui, S. Quetel-Weben, C.S. Sanmathi, R. Retoux, M. Gomina, Spark Plasma Sintering of n-Type Thermoelectric $Ca_{0.95}Sm_{0.05}MnO_3$, Journal of the American Ceramic Society, 94 (2011) 2608-2612.

[6] J.G. Noudem, A new process for lamellar texturing of thermoelectric $Ca_3Co_4O_9$ oxides by spark plasma sintering, Journal of the European Ceramic Society, 29 (2009) 2659-2663.

[7] J. Sun, G. Sun, W. Qu, Synthesis of dense NiZn ferrites by spark plasma sintering, Ceramics International, 28 (2002) 855–858.

[8] B.L. Shen, H. Kimura, A. Inoue, Fabrication of Fe-based glassy cores with high saturation magnetization and good soft magnetic properties by spark plasma sintering, Materials Science Forum, (2005) 3397-3400.

[9] W.S. Liu, B.P. Zhang, J.F. Li, H. L. Zhang, L. D. Zhao, Enhanced thermoelectric properties in CoSb3-xTex alloys prepared by mechanical alloying and spark plasma sintering, Journal of Applied Physics, 102 (2007) 103717.

[10] N. Frage, S. Cohen, S. Meir, S. Kalabukhov, M.P. Dariel, Spark plasma sintering (SPS) of transparent magnesium-aluminate spinel, Journal of Materials Science, 42 (2007) 3273-3275.

[11] T. Takeuchi, Y. Takeda, R. Funahashi, T. Aihara, M. Tabuchi, H. Kageyama, Rapid Preparation of Dense $(La_{0.9}Sr_{0.1})CrO_3$ Ceramics by Spark-Plasma Sintering, Journal of The Electrochemical Society, 147 (2000) 3979-3982.

[12] B. Zhang, Q. Zhao, A. Chang, Y. Li, Y. Liu, Y. Wu, Spark plasma sintering of $MgAl_2O_4$-$YCr_{0.5}Mn_{0.5}O_3$ composite NTC ceramics, Journal of the European Ceramic Society, 34 (2014) 2989-2995.

[13] M. Tokita, Mechanism of spark plasma sintering, in: Proceeding of NEDO International Symposium on Functionally Graded Materials, Japan, (1999) 22.

[14] L. Gao, H. Miyamoto, Spark plasma sintering technology, Journal of Inorganic Materials, 12 (1997) 129-133.

[15] M. Suárez, A. Fernández, J. Menéndez, R. Torrecillas, H. Kessel, J. Hennicke, R. Kirchner, T. Kessel, Challenges and Opportunities for Spark Plasma Sintering: A Key Technology for a New Generation of Materials, Intech-Open Access Publisher, (2013) 319-342.

[16] http://www.mfm.uni-jena.de/Forschung/FAST+_+SPS.html.

[17] X. Jin, A. Chang, H. Zhang, D. Zhang, Preparation and Microwave Sinterability of $Mn_{0.43}Ni_{0.9}CuFe_{0.67}O_4$ NTC Thermistor Materials by Pechini Method, Journal of Inorganic Materials, 24 (2009) 1013-1018.

[18] X. Jin, A. Chang, H. Zhang, D. Zhang, A Comparison Study of Sinterability and Electrical Properties for Microwave and Conventional Sintered $Mn_{0.43}Ni_{0.9}CuFe_{0.67}O_4$ Ceramics, Journal of Materials Science & Technology, 26 (2010) 344-350.

[19] A. Kamlo, J. Bernard, C. Lelievre, D. Houivet, Synthesis and NTC properties of $YCr_{1-x}Mn_xO_3$ ceramics sintered under nitrogen atmosphere, Journal of the European Ceramic Society, 31 (2011) 1457-1463.

[20] K. Park, J. Lee, The effect of ZnO content and sintering temperature on the electrical properties of Cu-containing $Mn_{1.95-x}Ni_{0.45}Co_{0.15}Cu_{0.45}Zn_xO_4$ ($0 \leq x \leq 0.3$) NTC thermistors, Journal of Alloys and Compounds, 475 (2009) 513-517.

[21] Y. Luo, X. Liu, G. Chen, Effect of Y_2O_3 addition on the electrical properties of $BaTiO_3$-based NTC thermistors, Materials Letters, 60 (2006) 1011-1013.

[22] K. Park, J. Lee, S.J. Kim, W.S. Seo, W.S. Cho, C.W. Lee, S. Nahm, The effect of Zn on the microstructure and electrical properties of $Mn_{1.17-x}Ni_{0.93}Co_{0.9}Zn_xO_4$ ($0 \leq x \leq 0.075$) NTC thermistors, Journal of Alloys and Compounds, 467 (2009) 310-316.

[23] D. Houivet, J. Bernard, J.M. Haussonne, High temperature NTC ceramic resistors (ambient-1000 ◦C), Journal of the European Ceramic Society, 24 (2004) 1237-1241.

[24] B. Zhang, Q. Zhao, A. Chang, H. Yan, Y. Wu, $MgAl_2O_4$-$LaCr_{0.5}Mn_{0.5}O_3$ composite ceramics for high temperature NTC thermistors, Journal of Materials Science: Materials in Electronics, 24 (2013) 4452-4456.

[25] M. Deepa, P.P. Rao, S. Sumi, A.N.P. Radhakrishnan, P. Koshy, New Negative Temperature Coefficient Ceramics in Ca-Ce-Nb-M-O (M=Mo or W) System, Journal of the American Ceramic Society, 93 (2010) 1576-1579.

[26] B. Zhang, Q. Zhao, A. Chang, J. Yao, P. Zhao, F. Guan, W. Kong, Complex impedance analysis of (Y_2O_3+CeO_2)–$YCr_{0.5}Mn_{0.5}O_3$ composite NTC ceramics, Journal of Alloys and Compounds, 512 (2012) 132-139.

[27] C. Nivot, J. Bernard, C. Lelievre, J.-M. Haussonne, D. Houivet, Moisture sensitivity of $YCr_{(1-x)}Mn_xO_3$ perovskites, Ceramics International, 36 (2010) 929-935.

[28] K. Fujiwara, S. Lee, N. Donnelly, T. Yamaguchi, C.A. Randall, Resistance Degradation in Y(Cr,Mn)O_3-Y_2O_3 Composite NTC Ceramics in Hostile Environments, Journal of the American Ceramic Society, 92 (2009) 2634-2641.

[29] T. Tachiwaki, Y. Kunifusa, M. Yoshinaka, K. Hirota, O. Yamaguchi, Formation, densification, and electrical conductivity of air-sinterable $Y(Cr_{1-x}Mg_x)O_3$ prepared by the hydrazine method, Materials Science and Engineering B, 86 (2001) 255-259.

[30] A. Durán, A. Arévalo-López, E. Castillo-Martínez, M. García-Guaderrama, E. Moran, M. Cruz, F. Fernández, M. Alario-Franco, Magneto-thermal and dielectric properties

of biferroic $YCrO_3$ prepared by combustion synthesis, Journal of Solid State Chemistry, 183 (2010) 1863-1871.

[31] T. Tachiwaki, Y. Kunifusa, M. Yoshinaka, K. Hirota, O. Yamaguchi, Formation, powder characterization and sintering of $YCrO_3$ prepared by a sol–gel technique using hydrazine, International Journal of Inorganic Materials, 3 (2001) 107-111.

[32] B. Zhang, Q. Zhao, A. Chang, Y. Li, Y. Liu, Y. Wu, Electrical conductivity anomaly and X-ray photoelectron spectroscopy investigation of $YCr_{1-x}Mn_xO_3$ negative temperature coefficient ceramics, Applied Physics Letters, 104 (2014) 102109.

[33] B. Zhang, Q. Zhao, A. Chang, Y. Liu, Y. Li, Y. Wu, Synthesis of $YCrO_3$ ceramics through a field-assisted sintering technique, Journal of Materials Science: Materials in Electronics, 25 (2014) 1400-1403.

[34] J. Yao, B. Zhang, J. Wang, A. Chang, G. Ji, P. Zhao, L. Zhao, Preparation of $Ni_{0.9}Mn_{2.1-x}Mg_xO_4$ ($0 \leq x \leq 0.3$) negative temperature coefficient ceramic materials by a rheological phase reaction method, Materials Letters, 112 (2013) 69-71.

[35] J. Wang, J. Zhang, Structural and electrical properties of $NiMg_xMn_{2-x}O_4$ NTC thermistors prepared by using sol-gel derived powders, Materials Science and Engineering: B, 176 (2011) 616-619.

[36] K. Park, Fabrication and Electrical Properties of Mn-Ni-Co-Cu-Si Oxides Negative Temperature Coefficient Thermistors, Journal of the American Ceramic Society, 88 (2005) 862-866.

Chapter 3

Al_2O_3 and Al_2O_3-ZrO_2 Fibers Obtained by Biotemplete with Low Thermal Conductivity

Tiago Delbrücke, Rogério A. Gouvêa,
Cristiane W. Raubach, Jose R. Jurado,
Faili C.T. Veiga, Sergio Cava, Mario L. Moreira and
Vânia C. Sousa

Additional information is available at the end of the chapter

http://dx.doi.org/10.5772/59011

1. Introduction

Certain porous materials have special properties and functions that cannot normally be obtained by conventional dense counterparts. Therefore, porous materials are now used in many applications such as final products and in various technological processes. Macroporous materials are used in various forms and compositions in everyday life; e.g. polymeric foams, packaging, lightweight aluminum structures in buildings, aircraft, and as a porous ceramic for water [1,2].

A growing number of applications that require advanced ceramics have appeared in recent decades, especially in environments where high temperatures, extended wear and corrosive environments are present. Such applications include the filtration of molten metals, high temperature insulation, support for catalytic reactions [3], filtration of particulates from exhaust gases of diesel engines and filtration of hot gases in various corrosive industrial processes, for example [4-6]. The advantages of using porous ceramic for these applications are generally a high melting point, suitable electronic properties, good corrosion resistance and wear resistance in combination with the characteristics acquired by the replacement of the solid material by voids in the component. Such characteristics include low thermal mass, low thermal conductivity, permeability control, high surface area, low density, high specific strength and a low dielectric constant [1,7]. These properties can be tailored for each specific application by controlling the composition and microstructure of the porous ceramic [8,9].

According to the type of application different microstructures and thus different preparation methods are required for porous ceramic bodies. For thermal or acoustic insulators materials with closed porosity are preferred, whereas membranes and filters require open pores with exactly defined pore size, for example. Are currently used some specific techniques to obtain pores with polymer and cotton for use in thermal insulation [10,11]. On the other hand, for application to environmental issues sodium hydroxide is used to pore formation [12].

This work uses a simple and versatile methodology to prepare sintered porous ceramic bodies obtained by replica method using organic fibers. The porous bodies of Al_2O_3 and Al_2O_3-ZrO_2 were synthesized through co-precipitation method [11] and replica [1,11] aiming to achieve a porous ceramic structure with thermal properties in agreement to the application of refractory materials.

2. Methodology

The Al_2O_3 fibers were prepared from an aluminum hydroxide slurry in which the organic fibers were dispersed and impregnated. The solvents were eliminated from the organic fibers through heat treatment and the remaining ceramic precursor was sintered to Al_2O_3.

The synthesis of aluminum hydroxide consisted of dissolving aluminum nitrate – $Al(NO_3)3.9H_2O$ (Synth) - in water; this solution was heated in a heating plate at 80°C. After its complete dilution ammonium hydroxide - NH_4OH (Vetec) - was added until pH = 9 was reached. The $Al(NO_3)$-NH_4OH ratio was 1:6 (in moles). Afterwards, anhydrous citric acid - $C_6H_8O_7$ (Synth) - was added until pH = 1 was reached, keeping a $Al(NO_3)$-$C_6H_8O_7$ molar ratio of 3:1. The addition of the citric acid causes the precipitation of aluminum nitrate in aluminum hydroxide - $Al(OH)_3$ - forming a slurry solution that was used to impregnate the organic fibers.

The organic fibers were impregnated in the aluminum hydroxide solution following the procedure of the replica method [1,11]. The organic fibers consisted of commercial cotton produced by Johnson & Johnson. After complete impregnation, the cotton fibers were pressed into a ceramic crucible to eliminate the surplus solution from the fibers and a green body with the shape of the recipient is obtained. These green bodies were placed on an electric oven (INTI, model FE-1300) for the calcination process, which occurred at 1200°C in air for 2 hours using a heating rate of 2°C/min. During this process there was a complete removal of the organic matter and α-alumina was formed from phase transition of aluminum hydroxide [13].

For the synthesis of zirconium hydroxide, initially, zirconium tetrachloride - $ZrCl_4$ (Synth) - was diluted in water. Next, this solution was heated in a heating plate at 80°C and after complete dilution ammonium hydroxide - NH_4OH-NH_4OH (Vetec) - was added until pH = 9 was reached keeping a $ZrCl_4$-NH_4OH molar ratio of 1:6. Afterwards, anhydrous citric acid - $C_6H_8O_7$ (Synth) - was added until pH = 1 was reached, keeping a $ZrCl_4$-$C_6H_8O_7$ molar ratio of 3:1. The addition of the citric acid causes the precipitation of zirconium tetrachloride in zirconium hydroxide - $Zr(OH)_4$ - forming a slurry solution that was used to impregnate the Al_2O_3 fibers.

http://dx.doi.org/10.5772/59011

The calcinated Al_2O_3 fibers were impregnated with the zirconium hydroxide slurry to form a structure of Al_2O_3 covered with ZrO_2. After impregnation, samples were sintered on the electric oven at 1200-1600°C in air for 4 hours using a heating rate of 2°C/min causing the transition of zirconium hydroxide to zirconia while the fibers are sintered [14].

2.1. Characterization

The crystalline phase was determined by X-ray diffraction (XRD) using a Shimadzu XRD-6000 diffractometer with CuKα; radiation at 40 kV and 40 mA with the patterns recorded in the 20 to 80 theta measuring range at a scan rate of 2°/min, at room temperature. Through micro-Raman was possible to verify the peaks corresponding to the symmetry of the ceramic phases; the analysis was performed in room temperature, using an wavelength of 514.5nm of an argon laser as exciting source. The energy was maintained in 15mW and a 50x lense was used. The spectra was registered through a monochromer T-64 Jobin-Yvon jointed to a CCD detector.

The morfology and structure of the pores created by the template of the metal oxides were analysed by FEG-SEM (Supra 35-VP, Carl Zeiss). Mean grain size was determined by using the intercept method.

A TG analysis of the porous alumina body was performed in a NETZSCH TG 209 F1 thermal analyzer using 10 mg samples heated to 25-900°C in air at a rate of 10°C/min.

Thermophysical properties are determined by the flash laser method [15-17] that allows to determine thermal diffusivity and specific heat of the sample. Specific heat and thermal diffusivity values were found through measurement of the increase in temperature of the opposite face of the material in the shape of a small disk, while the frontal face receives a strong energy flash by a laser. A laser with maximum power of 90 watts was applied. The irradiation time is within the order of 10 ms. The measurements were taken at room temperature in normal atmosphere.

The thermal diffusivity is calculated from the thickness of the sample and the required time for the temperature in the opposite face to reach 50% of the temperature in the laser incident face. Specific heat is determined by the density and thickness of the sample given the maximum temperature reached in the opposite face and the amount of heat received. Thermal conductivity is calculate by the product of the thermal diffusivity, specific heat and density, as shown in the equation 1 where K = thermal conductivity, α; = thermal diffusivity, ϱ; = density and C_p = specific heat.

$$K=\alpha \cdot \rho \cdot C_p \quad (1)$$

3. Results

Figure 1 shows the decomposition of organic fibers and aluminum hydroxide [18]. XRD patterns of samples in various temperatures suggest that phase changes occur at 200°C, 300°C,

400°C, 600°C, 700°C, 1100°C and 1600°C. The material phases agree with those phases obtained using the method of a solution precursor cation embedded in fibrous cotton organic matrix under high temperatures which profit from the reactive activity of decomposition and the reaction of fibers / aluminum hydroxide appears at 600°C but disappears at 700°C. At 700°C, the positions of all peaks agree with the positions of number 46-1212 of the JPCDS file (see Figure 1) which suggests that a completely crystallized Al_2O_3 product was obtained. A crystalline Al_2O_3 is obtained by the step-wise transition of $Al(OH)_3$ + cotton fibers between 300°C and 700°C. The results indicate that Al_2O_3 is the rhombohedral space group R-3C (167). Peaks were observed relative to α-Al_2O_3, and its crystalline peaks are identified in the X-ray diffractogram shown in Figure 1 as determined by the Scherrer equation [19], the average crystallite size is ~ 350 nm at 1600°C.

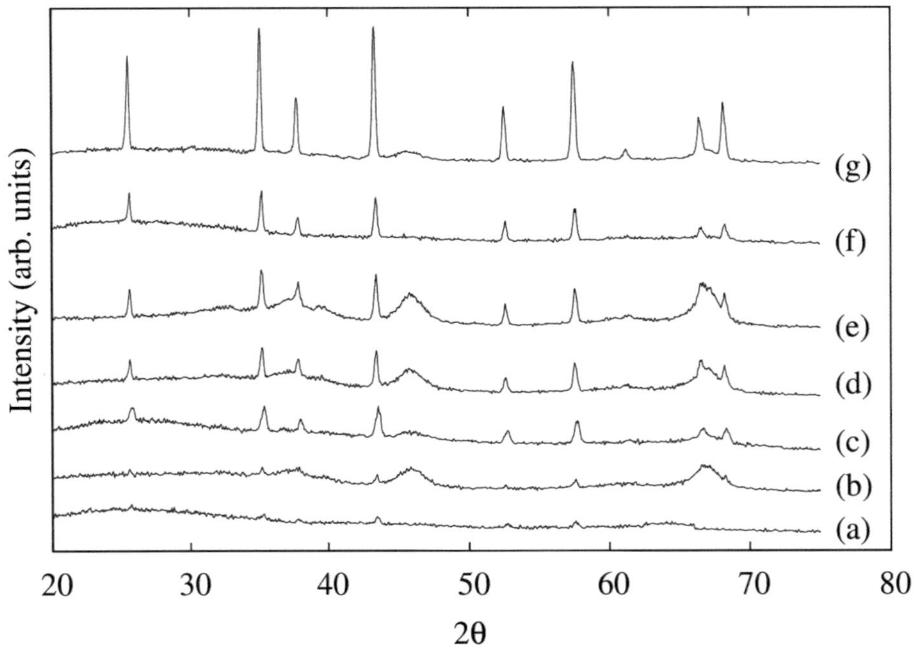

Figure 1. X-ray diffractogram of the Al_2O_3 porous bodies at temperatures of (a)200°C, (b)300°C, (c)400°C, (d)600°C, (e)700°C, (f)1100°C, (g)1600°C.

XRD of the sintered samples in different temperatures suggests the formation of alumina and zirconia phases in every temperature of sintering. The peaks for Al_2O_3-ZrO_2 are in agreement with those found in PDF (Powder Diffraction File) 53-559 file [20]; α-Al_2O_3 agrees with what is found in PDF 46-1212 file [21]; ZrO_2 agrees with the peaks in JPCDS 37-1484 file [22] (see Figure 2).

Raman is a powerful technique to detect the allotropic forms of zirconia [23]. According to the previous work of Popa et al [24], densified regions are caused by presence of monoclinic

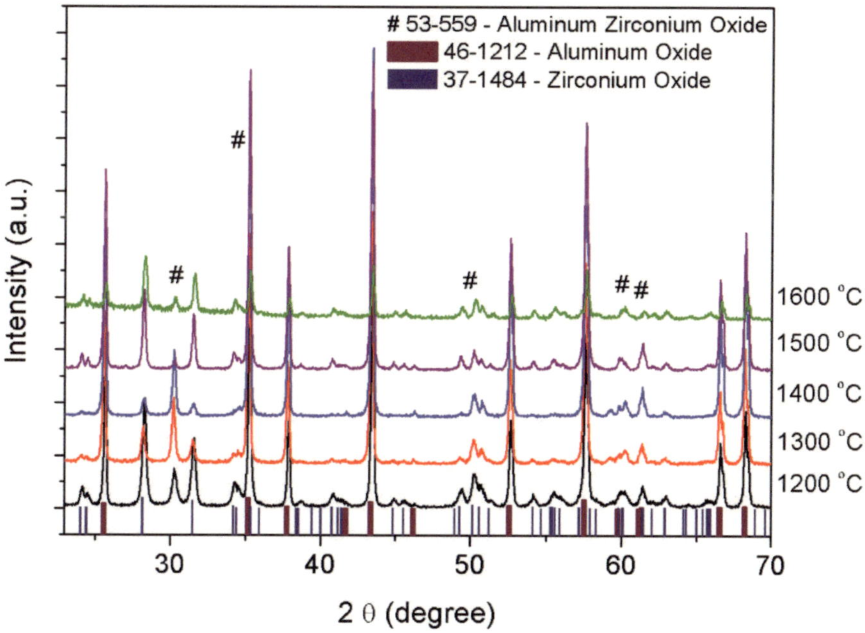

Figure 2. X-ray diffraction patterns of the Al_2O_3-ZrO_2 porous ceramics at several sintering temperatures.

zirconia. Figure 3 shows results regarding the measurements of micro-Raman spectroscopy dealing with wavelength radiation and vibration energies of the molecules. According to Raman results, vibration of Al-O bonds are related to the peaks described below: shows a peak at 378 cm-1, that may be considered polycrystalline α-Al_2O_3 [25]. For samples sintered above 1200°C there were significant spectroscopic bands in 410, 470, 605 and 630 cm^{-1}, which are identified as absorption bands characteristic of α-Al_2O_3 [26-28]. These results are agreement with the study conducted by Cava et al, in which spectroscopic bands of α-Al_2O_3 are present in temperatures above 1000°C [13].

Figure 4 shows the morphological profile of the Al_2O_3 and Al_2O_3-ZrO_2 porous body with an increase of 5000x where the samples are not similar. Densification and grain growth are two inversely proportional properties during the sintering process. The high porosity and low densification shown in SEM images of the Al_2O_3 porous bodies (Figure 4a) relates to a heating rate of 2°C/min which was used during the sintering process [29]. SEM micrographs (Figure 4b) show the presence of densified regions of zirconia; therefore, micro-Raman was used to identify the presence of monoclinic zirconia in these regions. Vibration of Zr-O bonds are related to the peaks on the spectroscopic bands of monoclinic zirconia were identified in frequencies of 505, 534, 550 cm^{-1} [24,30], peaks shown in temperatures of 1300-1600°C. Only a small amount of monoclinic zirconia is present; however it can be suggested that this amount was enough to densify the regions of porous ceramics agglomerating around grains [24]. Tetragonal zirconia was clearly detected in the peaks 260, 300, 323 and 340 cm^{-1} [24] increasing their intensity above 1300°C. It was found only two well defined peaks at 217 and 745

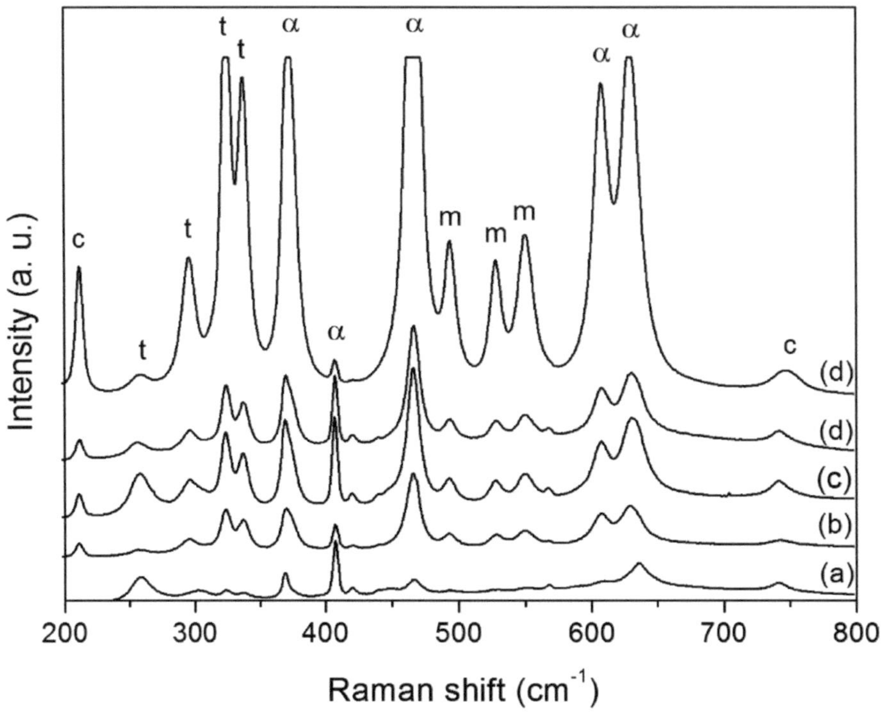

Figure 3. Micro-Raman spectra of the Al_2O_3-ZrO_2 porous ceramics at differents temperatures of: (a) 1200°C, (b) 1300°C, (c) 1400°C, (d) 1500°C and (e) 1600°C. α = α-alumina, c = cristobalite zirconia, t = tetragonal zirconia, m = monoclinic zirconia.

corresponding to cristobalite zirconia between temperatures of 1200-1600°C. The crystalline phase of Al_2O_3-ZrO_2 shown in Figure 2 has a cubic structure [20], not being detected by Raman.

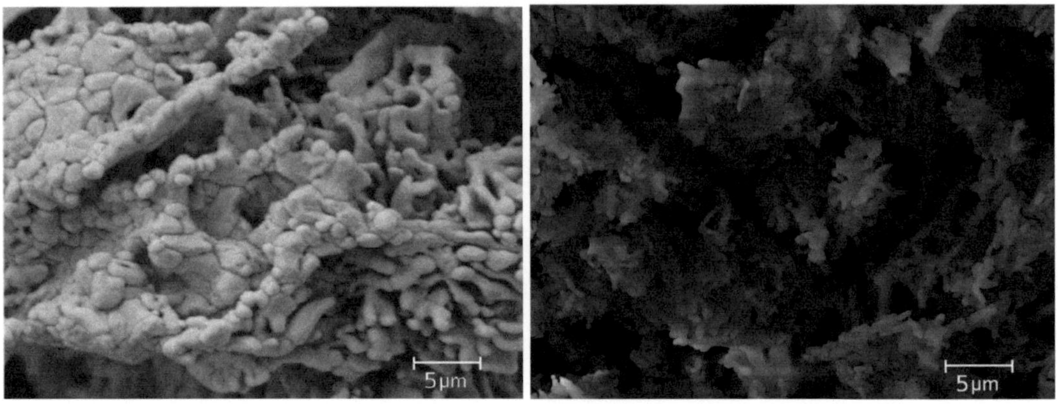

Figure 4. SEM micrographs of Al_2O_3 and Al_2O_3-ZrO_2 porous bodies at temperature of 1600°C at increasing of 5000x.

http://dx.doi.org/10.5772/59011

The method of embedding a cotton fiber in the organic matrix is based on the impregnation of a cellular structure with a ceramic suspension or solution precursor ceramic to produce a macroporous ceramic which has the same morphology as the original porous material (cotton) as illustrated in SEM images of Figure 4. Many cellular structures can be used as templates to produce macroporous ceramic embedding techniques for organic matrices.

The process of manufacturing ceramic fibers using cotton as a template by employing the embedding of the organic matrix [1] can be divided into two stages: the formation of the Al_2O_3 fiber morphology and the removal of the cotton used as a template.

The TG illustrated in Figure 5 shows the weight loss along with images of samples at temperatures ccurring in the formation of pore bodies at a temperature of 900°C; a gradual weight loss of weight is evident during the process. The sample weight loss was 82.5% up to 650°C where it terminated the decomposition of the carbonaceous mass with no significant loss of mass after the complete elimination of the carbonaceous mass.

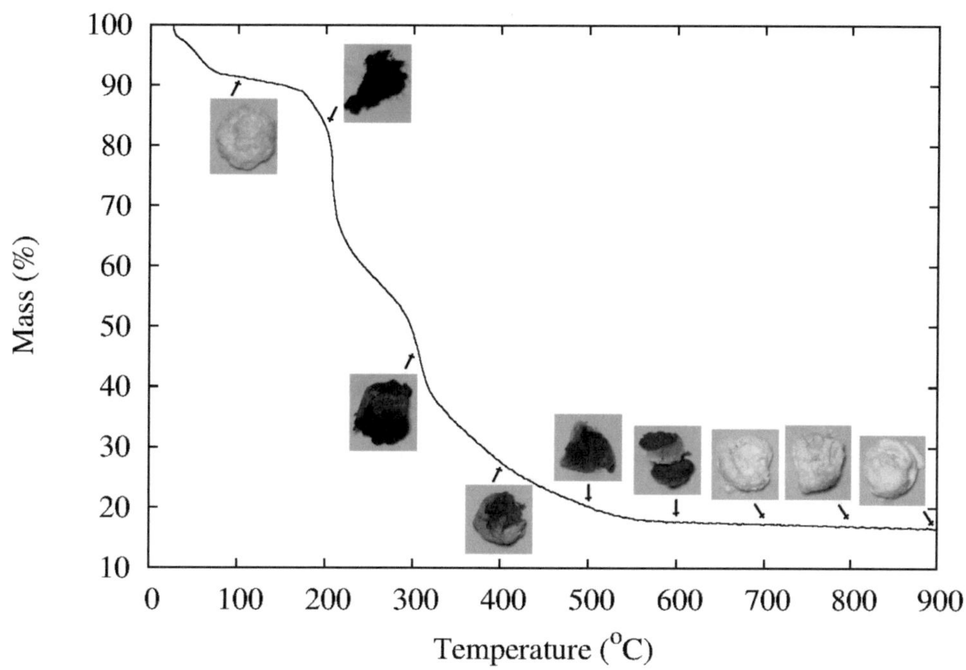

Figure 5. Thermogravimetric analysis of the Al_2O_3 porous bodies.

Al_2O_3 fibers are formed during the calcination process. After the formation of Al_2O_3 fibers, cotton models are removed through the decomposition process (see Figure 5) [31] which begins below 300°C, and terminates close to 700°C.

The porous body at 100°C has a weight loss of 9% which indicates that only evaporation solvents are used in the TG analysis. A weight loss of 8% over a primary loss of 9% at 200°C indicates the evaporation of water in the sample and the beginning of the organic material evaporation. An additional weight loss of 55% at 400°C indicates the organic matter evaporation from the cotton fiber which forms the porous body [31]. A mass loss of 11% and a final weight of 83% in relation to the green body which indicates the total organic matter evaporation and the formation of the defined porous body.

Figure 4b shows morphological features of the porous ceramic of Al_2O_3-ZrO_2, in which grain growth was possible due to the presence of ZrO_2, causing the formation of diffusion barriers to control the growth and formation of the grain[32]. Furthermore, it was possible to verify the solid phase sintering, where the densification and grain growth are controlled by the diffusion across the grain boundary [33]. Therefore, it is reasonable to affirm that the formation of porosity in the structure is beneficial to decrease the thermal conductivity of the sample.

Table 1 illustrates physical and geometric properties of Al_2O_3 and Al_2O_3-ZrO_2 porous bodies.

Sample	Surface area (m^2/g)	Pore Volume (cm^3/g)	Pore Diameter (Å)	Porosity (%)
Al_2O_3	14.33	0.01	42.47	40.7
Al_2O_3-ZrO_2	8.45	0.006	4.19	77.9

Table 1. Physical and geometrical properties of the Al_2O_3 and Al_2O_3-ZrO_2 porous ceramics.

For the porosity calculation by the average of 3 samples, the Al_2O_3 theoretical density value was assumed to be 3940 $kg.m^{-3}$; the actual density was 2336 $kg.m^{-3}$. A porosity average value of 40.7% and 77.9% was obtained for the Al_2O_3 and Al_2O_3-ZrO_2 porous, respectively.

Table 2 shows the results of thermophysical property determinations of Al_2O_3 and Al_2O_3-ZrO_2 porous bodies.

Sample	a ($10^6 m^2 s^{-1}$)	($Kg.m^{-3}$)	C_p ($J.Kg^{-1}.K^{-1}$)	K ($W.m^{-1}.K^{-1}$)
Al_2O_3	1.24	2336	561.8	1.63
Al_2O_3-ZrO_2	1.09	2696	547.3	1.61

Table 2. Analysis of thermal conductivity by laser flash method of the Al_2O_3 and Al_2O_3-ZrO_2 porous bodies.

In the literature very few data about the thermal conductivity of the Al_2O_3 and ZrO_2 system is available. However the effect of the porosity to reduce the thermal conductivity of ceramic materials is a well recognized phenomenon, that has been widely applied for Al_2O_3 and ZrO_2 [3,10,34]. In addition to the absolute value of the porosity, the interconnection of grain size and pore shape have a significant influence on the final thermal conductivity. The high purity Al_2O_3 with no pores and with average grain size ~1 μm shows a thermal conductivity of approximately 33 $W.m^{-1}.K^{-1}$ at room temperature. High purity ZrO_2 without porosity and average grain size of ~1 μm has a thermal conductivity of approximately 3.3 $W.m^{-1}.K^{-1}$ [35] at room temperature. Different studies show that thermal conductivity is independent of the grain size [36,37].

Some studies have been reported in the literature on the effect of porosity on reducing the thermal conductivity of solids, especially the porosity of Al_2O_3 [3,10]. In addition to its absolute value, the grain size and porous interconnectivity and shape have a significant influence on the final thermal conductivity. Porosity-free high-purity Al_2O_3 with a grain size ~1 μm has a thermal conductivity of approximately 33 $W.m^{-1}.K^{-1}$ at room temperature [35] which is very high when compared to the results obtained for porous Al_2O_3 reported by B. Nait-Ali et al [10] and Z. Zivcova et al [3]. Note that for the relative thermal conductivity the grain size dependence of thermal conductivity [36,37] is irrelevant, since it is cancelled out by taking the conductivity ratio [3].

Nait-Ali and co-workers [10] conducted a research on the Al_2O_3-ZrO_2 system relating thermal conductivity to porosity. Their samples were sintered at 1400°C and pores were generated by a pore-forming polymer. Results showed that commercial Al_2O_3 with 40% of porosity presented a thermal conductivity of 9 $W.m^{-1}.K^{-1}$ and average grain size of 0.5 μm; for commercial ZrO_2, thermal conductivity was 0.9 $W.m^{-1}.K^{-1}$ for 37% of porosity and average grain size of 0.1 μm.

The obtained Al_2O_3 porous bodies sintered at 1600°C have a thermal conductivity of 1.63 $W.m^{-1}.K^{-1}$ with a porosity of 40.71% and average grains size 0.55 μm, using cotton pore-forming agent and alumina obtained by a phase transition. Correlating these values and methods with literature data shows that Al_2O_3 porous bodies have high refractory properties from the combination of factors such as the synthesis method, grain size and porosity, when compared to the thermal conductivity of alumina bodies analyzed at temperatures as high as 1000°C [38].

Bansal and co-workers [39], using samples of commercial ZrO_2-Al_2O_3 sintered at 1000°C with density around 99% with average grain size of ~1 μm showed a thermal conductivity of 6.9 $W.m^{-1}.K^{-1}$. The porous Al_2O_3-ZrO_2 fibers obtained in this work were sintered at 1600°C and the calculated thermal conductivity was 1.61 K($W.m^{-1}.K^{-1}$) with a porosity of 77.9% and an average grain size of ~1 μm. The cotton replicated fibers of Al_2O_3-ZrO_2 sintered at 1600°C presented very low thermal conductivity compared to other works using different processes of pore formation.

4. Conclusions

Al_2O_3 porous bodies composed of ceramic fibers were successfully obtained by the embedded fibrous organic matrix method with cotton as a template. SEM at different temperatures during heat treatment along with thermogravimetric analysis data indicates a step-by-step method for the complete formation of the ceramic fiber porous body. The sintering temperature, low heating rate and the use of cotton as template had a strong effect on the surface area, pore size and distribution of the synthesized fibers. Thermal conductivity data show excellent results when compared to the literature, due to the direct influence of the organic template as a shape-model and the efficient method of synthesis. The results show that the Al_2O_3 and Al_2O_3-ZrO_2 porous body are an excellent thermal insulator with direct application for refractories. A higher porosity and lower densification of the porous body is made possible with the addition of ZrO_2 to the Al_2O_3 matrix. However, there was no difference in thermal conductivity due to the characteristic values of density, specific heat and microstructure observed in both materials.

Author details

Tiago Delbrücke[1], Rogério A. Gouvêa[1], Cristiane W. Raubach[1], Jose R. Jurado[1], Faili C.T. Veiga[1], Sergio Cava[1], Mario L. Moreira[2] and Vânia C. Sousa[3]

1 Graduate Program in Science and Materials Engineering, Technology Development Center, Federal University of Pelotas, Pelotas, Brazil

2 Institute of Physics and Mathematics, Federal University of Pelotas, Pelotas, Brazil

3 Engineering Materials Department, Federal University of Rio Grande do Sul, Porto Alegre, Brazil

References

[1] Studart, A., Gonzenbach, U., Tervoort, E. & Gauckler, L. (2006). Processing routes to macroporous ceramics: a review, J. Am. Ceram. Soc. 89(6): 1771-1789.

[2] Ohji, T. & Fukushima, M. (2012). Macro-porous ceramics: processing and properties, Int. Mater. Rev. 57(2): 115-131.

[3] Zivcova, Z., Gregorova, E., Pabst, W., Smith, D., Michot, A. & Poulier, C. (2009). Thermal conductivity of porous alumina ceramics prepared using starch as a pore-forming agent, J. Eur. Ceram. Soc. 29(3): 347-353.

[4] Okada, K., Shimizu, M., Isobe, T., Kameshima, Y., Sakai, M., Nakajima, A. & Kurata, T. (2010). Characteristics of microbubbles generated by porous mullite ceramics prepared by an extrusion method using organic fibers as the pore former, J. Eur. Ceram. Soc. 30(6): 1245-1251.

[5] Okada, K., Uchiyama, S., Isobe, T., Kameshima, Y., Nakajima, A. & Kurata, T. (2009). Capillary rise properties of porous mullite ceramics prepared by an extrusion method using organic fibers as the pore former, J. Eur. Ceram. Soc. 29(12): 2491-2497.

[6] Okada, K., Imase, A., Isobe, T. & Nakajima, A. (2011). Capillary rise properties of porous geopolymers prepared by an extrusion method using polylactic acid (PLA) fibers as the pore formers, J. Eur. Ceram. Soc. 31(4): 461-467.

[7] Schacht, C. (2004). Refractories handbook, Vol. 178, CRC.

[8] Dong, Q., Su, H., Xu, J., Zhang, D. & Wang, R. (2007). Synthesis of biomorphic ZnO interwoven microfibers using eggshell membrane as the biotemplate, Mater. Lett. 61(13): 2714-2717.

[9] Dong-Dong, W., Gang, W., Xiao-Fei, S., Ya-Ping, L., Shu-Qiang, D. & Hong-Xia, L. (2012). Fabrication of nanoporous mullite ceramics, Chinese J. Inorg. Chem. 28(3): 491-494.

[10] Nait-Ali, B., Haberko, K., Vesteghem, H., Absi, J. & Smith, D. (2007). Preparation and thermal conductivity characterisation of highly porous ceramics: Comparison between experimental results, analytical calculations and numerical simulations, J. Eur. Ceram. Soc. 27(2-3): 1345-1350.

[11] Delbrücke, T., Gouvea, R. A., Moreira, M. L., Raubach, C. W., Varela, J. A., Longo, E., Gonçalves, M. R. & Cava, S. (2012). Sintering of porous alumina obtained by biotemplate fibers for low thermal conductivity applications, Journal of the European Ceramic Society.

[12] Bento, A. C., Kubaski, E. T., Sequinel, T., Pianaro, S. A., Varela, J. A. & Tebcherani, S. M. (2013). Glass foam of macroporosity using glass waste and sodium hydroxide as the foaming agent, Ceramics International 39(3): 2423-2430.

[13] Cava, S., Tebcherani, S., Souza, I., Pianaro, S., Paskocimas, C., Longo, E. & Varela, J. (2007). Structural characterization of phase transition of Al_2O_3 nanopowders obtained by polymeric precursor method, Mater. Chem. Phys. 103(2-3): 394-399.

[14] Ussui, V., Leitão, F., Yamagata, C., Menezes, C. A., Lazar, D. R. & Paschoal, J. O. (2003). Synthesis of ZrO_2-based ceramics for applications in sofc., Materials science forum, Vol. 416, pp. 681-686.

[15] Parker, W., Jenkins, R., Butler, C. & Abbott, G. (1961). Flash method of determining thermal diffusivity, heat capacity, and thermal conductivity, Jpn. J. Appl. Phys. 32(9): 1679-1684.

[16] Taylor, R. (1979). Heat-pulse thermal diffusivity measurements, High Temp.- High Pressures 11(1): 43-58.

[17] Ferreira, R., Miranda, O., Dutra Neto, A., Grossi, P., Martins, G., Reis, S., Alencar, D., Soares Filho, J., Lopes, C. & Pinho, M. (2002). Implantação no CDTN de laboratório de medição de propriedades termofísicas de combustíveis nucleares e materiais através do método flash laser, 2002 International Nuclear Atlantic Conference-INAC 2002, pp. 11-16.

[18] Deng, Z., Fukasawa, T., Ando, M., Zhang, G. & Ohji, T. (2001). High-surface-area alumina ceramics fabricated by the decomposition of al (oh) 3, J. Am. Ceram. Soc. 84(3): 485-491.

[19] Cullity, B. & Stock, S. (1972). Elements of X-ray Diffraction, Vol. 170, Prentice Hall.

[20] Kimmel, G., Zabicky, J., Goncharov, E., Mogilyanski, D., Venkert, A., Bruckental, Y. & Yeshurun, Y. (2006). Formation and characterization of nanocrystalline binary oxides of yttrium and rare earths metals, Journal of alloys and compounds 423(1-2): 102-106.

[21] Maslen, E., Streltsov, V., Streltsova, N., Ishizawa, N. & Satow, Y. (1993). Synchrotron x-ray study of the electron density in-Al_2O_3, Acta Crystallographica Section B: Structural Science 49(6): 973-980.

[22] McMurdie, H. F., Morris, M. C., Evans, E. H., Paretzkin, B., Wong-Ng, W. & Hubbard, C. R. (1986). Methods of producing standard x-ray diffraction powder patterns, Powder Diffraction 1(01): 40-43.

[23] Kakihana, M., Yashima, M., Yoshimura, M., Borjesson, L. & Mikael, K. (1993). Application of raman spectroscopy to phase characterization of ceramic hightc superconductors and zirconia related materials, Resesarch Trends (1): 261-311.

[24] Popa, M., Kakihana, M., Yoshimura, M. & Calderón-Moreno, J. M. (2006). Zircon formation from amorphous powder and melt in the silica-rich region of the alumina-silicazirconia system, Journal of non-crystalline solids 352(52): 5663-5669.

[25] Watson, G., Daniels, W. & Wang, C. (1981). Measurements of raman intensities and pressure dependence of phonon frequencies in sapphire, Journal of Applied Physics 52(2): 956-958.

[26] Nagabhushana, K., Lakshminarasappa, B. & Singh, F. (2009). Photoluminescence and raman studies in swift heavy ion irradiated polycrystalline aluminum oxide, Bulletin of Materials Science 32(5): 515-519.

[27] Boumaza, A. & Djelloul, A. (2010). Estimation of the intrinsic stresses in α-alumina in relation with its elaboration mode, Journal of Solid State Chemistry 183(5): 1063-1070.

[28] Mariotto, G., Cazzanelli, E., Carturan, G., Di Maggio, R. & Scardi, P. (1990). Raman and x-ray diffraction study of boehmite gels and their transformation to α- or β-alumina, Journal of Solid State Chemistry 86(2): 263-274.

[29] Zhou, Y., Hirao, K., Yamauchi, Y. & Kanzaki, S. (2004). Densification and grain growth in pulse electric current sintering of alumina, J. Eur. Ceram. Soc. 24(12): 3465-3470.

[30] Boullosa-Eiras, S., Vanhaecke, E., Zhao, T., Chen, D. & Holmen, A. (2011). Raman spectroscopy and x-ray diffraction study of the phase transformation of ZrO_2-Al_2O_3 and CeO_2-Al_2O_3 nanocomposites, Catalysis Today 166(1): 10-17.

[31] Fan, T., Sun, B., Gu, J., Zhang, D. & Lau, L. (2005). Biomorphic Al_2O_3 fibers synthesized using cotton as bio-templates, Scr. Mater. 53(8): 893-897.

[32] Caruso, F., Caruso, R. & Möhwald, H. (1998). Nanoengineering of inorganic and hybrid hollow spheres by colloidal templating, Science 282(5391): 1111.

[33] Bernard-Granger, G. & Guizard, C. (2008). New relationships between relative density and grain size during solid-state sintering of ceramic powders, Acta Materialia 56(20): 6273-6282.

[34] Nait-Ali, B., Haberko, K., Vesteghem, H., Absi, J. & Smith, D. (2006). Thermal conductivity of highly porous zirconia, Journal of the European Ceramic Society 26(16): 3567-3574.

[35] Pabst, W. & Gregorová, E. (2007). Effective thermal and thermoelastic properties of alumina, zirconia and alumina-zirconia composite ceramics, New Developments in Materials Science Research pp. 77-137.

[36] Smith, D., Fayette, S., Grandjean, S., Martin, C., Telle, R. & Tonnessen, T. (2003). Thermal resistance of grain boundaries in alumina ceramics and refractories, Journal of the American Ceramic Society 86(1): 105-111.

[37] Smith, D., Grandjean, S., Absi, J., Kadiebu, S. & Fayette, S. (2003). Grain-boundary thermal resistance in polycrystalline oxides: alumina, tin oxide, and magnesia, High Temperatures-High Pressures 35(1): 93-100.

[38] Taylor, R. & Dos Santos, W. (1993). Effect of porosity on the thermal conductivity of alumina, High Temp.- High Pressures 25: 89-98.

[39] Bansal, N. & Zhu, D. (2005). Thermal conductivity of zirconia-alumina composites, Ceramics international 31(7): 911-916.

Section 2

Sintering with Additives

Chapter 4

The New Generation of Diamond Wheels with Vitrified (Ceramic) Bonds

Barbara Staniewicz-Brudnik, Elżbieta Bączek and Grzegorz Skrabalak

Additional information is available at the end of the chapter

http://dx.doi.org/10.5772/59503

1. Introduction

Modern, composite materials: non-ferrous metal alloys reinforced with hard particles, wood composites, plastic composites reinforced with glass fibers, nickel super-alloys (Hastalloy, Waspalloy, Inconel 718, Udimet720) and titanium Ti6Al4V, TiAl) [1-5] requires the development of technology taking into account abrasive tools with new binders [6-9].It is emphasized that provided affective work tools in the optimal adjustment of the characteristics of grinding machines, machining parameters, the characteristic of tools and a method of cooling and dressing. Typical bonded abrasive tool includes: super hard grains (diamond cubic boron nitride, mono-or microcrystalline microstructure), filler(corundum, silicon carbide, boron nitride etc.), binder (sintered metal, electroplated, resin, ceramic, hybrid), modifiers (homogenizers, greases), body (metal, ceramic, polymer). Currently as bonded abrasive products are mainly used synthetic diamond or cubic boron nitride grains. Significant diamond grains producers were shown in table 1 [10]

Producer	Trade name of diamond	Grains morphology	Application
Sandvic Hyperion, USA	MBG 600	Irregular, sharp, friable	Coarse and fine grinding ceramics, glass and precious
	MBG 640	Cubooctahedral crystals with high impact and fracture strength, free cutting capability	

Producer	Trade name of diamond	Grains morphology	Application
	RVG	Irregular shapes, medium friable crystals	Precision grinding and finishing of tungsten carbide and carbide, grinding of
	RVG 800	Irregular shape, superior free capabilities due to controlled micro fracturing mode of crystals	
	RBI - uncoated	Irregular shape, medium friable crystals	grinding of tungsten carbides and PCD inserts
	SPR	Uncoated economic grade grain, the friability of SPR enables controlled diamond fracturing, wheel self-sharpening and free cutting action	
Element Six, USA	PDA 657		It is suited for used in less applications where sharp cutting characteristic are important
	PDA 321 PDA 311 PDA 211	microchipping structure, friable	Grinding of cermets and elements of technical ceramic
Gems Superabrasives, China	GRD 10 GRD 20	Multi crystal shape, consisting of many micro crystals with slight friability enable excellent self- sharpness	Grinding of soft Stones cutting,, sharpening, grinding ceramics and rubber
Lands, USA	LS 070	Ultra friable, irregular shape	Specialized grinding benefiting from free cutting crystals
	LS 100	Friable, irregular shape	
	LS 600F	Uniform, blocky friable	Grinding lapping, polishing of cemented carbide, special steels, glass, natural stone
Kingray New Materials & Science Technology Co., China	JRI – typ A JRI – typ B	irregular shape	Grinding of nonferrous hard alloys
Changha 3 better ultra HARD MATERIALS, China	SMD 690	Premium grade grain with high strength and great thermal stability	
	MBD 6 MBD 4	Sharp crystal shape	Grinding stones ,ceramics, polishing of hard materials optical glass grinding
	RVD	Irregular crystal shape preferably self-sharpness	Grinding of hard alloy, non-ferrous metal, polishing of natural diamond
	RVE	Friable and strong self-sharpness, policrystallic shape, low hardness	Fine processing ceramics, optical glasses, grinding, polishing of

Producer	Trade name of diamond	Grains morphology	Application
			nonferrous metals and nonmetallic materials, PCD, carbide, granite
SF Diamond, China	SFD 10	Irregular shape, rough surface low strength	For grinding tools, which work with low load
	SFD 20 SFD 30	Irregular crystal shape, rough surface, low strength	
	SFD 40 SFD 50	Comparatively regular crystal shape, medium strength	For grinding wheels, which work with medium and low load
	SDM-c	Dark green color, medium tough, micro-friable multinano crystal, irregular shape, perfect self-sharpen ability	
The Institute of Super Hard Materials Kiev Ukraine	AC 4 AC 6	Irregular shape, friable, low strength	Grinding of brittle materials (ceramic, hard alloys, glass)
	AC 15 AC 30	Irregular shape, less friable, medium strength	
	AC 100 ACR	Comparatively regular crystal shape, medium strength	

Table 1. Significant producers of diamond grains

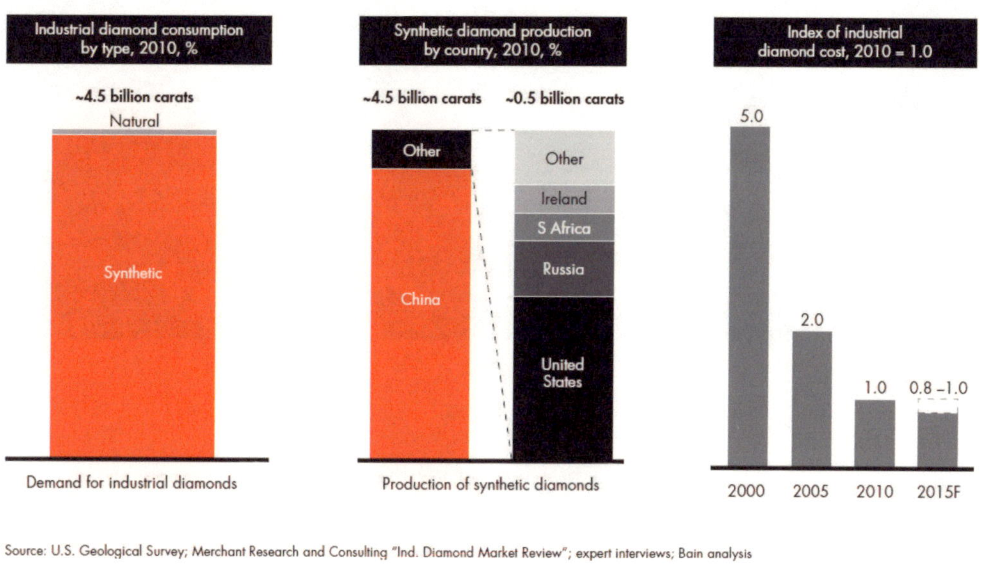

Figure 1. Synthetic industrial diamonds production [1]

In general, monocrystalline grains, used for tools with vitrified bond are assigned for grinding of flat and shaped surfaces, for example grinding of internal surfaces of Titan alloys and sharpening (fig. 2). Microcrystalline grains are complexes consisting of collection of small microcrystals with sizes ranging from one to a few microns, characterized by a higher mechanical strength, and higher ductility of monocrystalline grains (fig. 3)

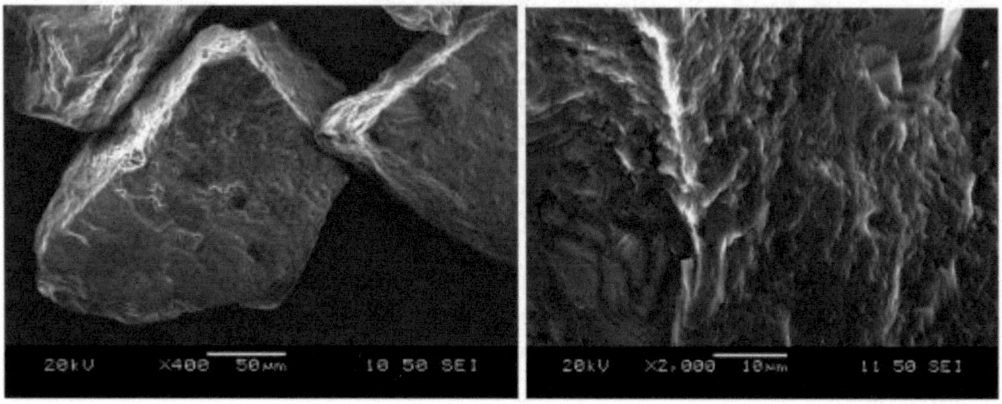

Figure 2. Microcrystalline diamond grains magn.: a) 400x, b) 2000x

Figure 3. Monocrystalline diamond grains magn.: a) 200x, b) 2000

The microcrystalline grains are recommended for grinding operations, where surface quality is the main processing criterion. The fillers can be alumina, silicon carbide, tungsten, zirconium silicate etc. It is assumed that the filler, introduced into the binder, protects the seed prior to the dynamic action of chips (typically having a higher temperature) and that increases the strength properties of the sinter, the thermal resistance and wear resistance, also taking part in the grinding process. The binder is a factor connecting the embankment of grain in the wheel causes the maintenance work until the grains have sharp edges, thereby induced on the self-sharpening effect of the grain. The efficiency of binders application provide features such as adequate strength of fixing force of particles in the binder, wear resistance, the possibility of sintering of the tools below temperature of graphitization (diamond) or active oxidation (CBN), having long thermal conductivity allowing for intense heat removing from the grinding zone without accumulation, the coefficient of thermal expansion close or identical with the coefficient of thermal expansion of the abrasive grain, proper hardness and strength, no reaction with work-piece material [6-8, 11-13]. Vitrified binders are primarily glass or devitrificate from the multicomponent system, received from synthetic chemicals with very high purity [6, 9].

They are divided into a low-melting binder (sintering temperature up to 730° C), medium (sintering temperature up to 880°C) and a high-melting (sintering temperature above 900°C). The most characteristic systems are borosilicate glasses modified with various oxides (Na_20, Li_20), calcium-silicate glasses, modified by V_20_5, Fe_20_3 or P_20_5, lead-borosilicate devitrificates, barium-borosilicate with the addition of bismuth oxide. The most famous research centers dealing with binders for super hard tools including vitrified bond are located in Worceaster (GE company-USA), Aachen(RWTH-Germany), Sankt Petersburg (Ilyich company –Russia), Kiev (Institute of Super hard Materials –Ukraine), Zhengzhou (Zhengzhou Concern Hongtuo Superabrasive Products Co.-China). The Institute of Advanced Manufacturing Technology is conducting researches aiming at obtaining ceramic binders for super abrasives tools with controlled physical and mechanical properties and testing tools containing these binders. This publication shows the results of research of new vitrified (ceramic) binders and testing of grinding tools with these binders for the latest generation of composite BNDCC (boron nitride dispersive in cemented carbide) of various grit size of cubic boron nitride grains.

2. Experimental section

2.1. Part I Preparation and study of some physical and mechanical properties of glasses

2.1.1. The starting materials and methods of research

The study selected five variants of glasses from the $Si0_2$-Al_20_3-B_20_3-Na_20-Ba0 system. The following were used as starting materials:

- silicon dioxde -$Si0_2$ pure
- hydrated aluminum nitrate Al $(N0_3)_3$ x$9H_20$ of high purity or aluminum hydroxide Al $(0H)_3$
- barium carbonate $BaC0_3$-pure
- boric acid H_3B0_3-pure
- sodium carbonate Na_2C0_3-pure

After accurate grinding, sieving through a 0.63mm sieve and mixed, raw materials were placed in corundum crucibles and heated at a temperature above 1,350° C. After fritting (hot melt glass pouring into cold water) and once again the glass was milled and sieved. All glass frits were completely transparent with bluish color. On the received materials the following tests were carried out:

- the density of helium, on Accu Pyc II 1340V10 helium pycnometer using 5 parallel samples;
- the X-ray diffractometer of Panalytical Empiriam with copper lamp in the range of 2 theta angle of 5 to 90 degrees;
- calculation of thermodynamic stability using algorithm VCS for sets of glasses in temperatures of 700, 900, 1300 and 1500° C on two assumptions: the total miscibility of liquid and vapor phases or total immiscibility liquid and vapor phases;
- the reactivity of raw materials of glasses and the melt glasses using differential scanning calorimetry (DSC-device STA-449 F3 Jupiter)
- the wettability of glasses to the substrate of silicon carbide or sintered alumina (cubitron) on the high-temperature microscope of Leitz -Watzler type by sessile-drop method.
- microscopic observation (SEM)of transverse specimens after testing the wettability
- the Young's modulus of glass using a flaw with the head broadband
- the three-point bending strength of trabecular glass, describing the work of destruction by using a machine Zwick Roell Z2.5

2.1.2. Results and discussion

Test glass belonged to the group of light, characterized by a low density, ranging between 2.34 (W4), and 2.69 g/cm^3 (Ba23bis). It was evident that with the reduction of silicon dioxide content in the glass, decreasing the density of the glass increased A slight increase in the content of barium oxide, the constant of silicon dioxide and reduced alumina content did not affect significantly the increase in density. The results are summarized in Table 2

The variant of glasses	Density [g/cm^3]
W1	2,4548
W2	2,5347
W3	2,4320
W4	2,3426
Ba 23 bis	2,6912

Table 2. The density of the tested glasses

2.1.3. Calculations of thermodynamic stability of the glasses by algorithm VCS

Chemical stability of the connections between components of the glasses was determined by calculation of the thermodynamic potential of components by VCS algorithm, which takes into account the probable stability of the reaction products. Equilibria calculations were performed for the four variants of the glasses of the SiO_2-Al_2O_3-BaO-B_2O_3-Na_2O system in temperatures 700° C, 900° C, 1,350° C, 1500° C at atmospheric pressure 1013 hPa. The molar ratios of the components were adopted taking into account their actual values. It was assumed that the component of glass within these ranges of temperatures may occur: in a multicomponent gas phase and condensed phase pure (liquid and solid). It is not known whether the active phase pure liquid form (that are immiscible with each other) or a liquid phase formed of unlimited miscibility, therefore, the calculation was performed by two assumptions out. Of the approximately 100 likely stable compounds there were 10, of which the solid 3. Summary of solid stable compounds glass as an example of option 2 are shown in Tables 3 and 4.

Name of compounds in solid phase	700° C	900° C	1350 ° C	1500 ° C
$NaAlSi_3O_8$	0,9096	0,92326	–	–
$Al_4B_2O_9$	0,0772	0,0430	–	–
$Al_6BSi_2O_{13}$	–	–	0,15674	0,15674
$BaSi_2O_5$	–	–	0,08201	0,08201
SiO_2	–	–	2,4153	2,4153

Table 3. Calculation of the thermodynamic stability of glass precursors, option 1 assumption 1

According with the first assumption at high temperatures (1350, 1500° C) there should be stable three compounds: an aluminum borosilicate ($Al_6BSi_2O_{13}$), barium silicate ($BaSi_2O_5$) and silicon dioxide (SiO_2). In the second assumption the second barium silicate ($BaSi_2O_5$) presented only one. Similarly were in the other glasses. The verification of the existence of these compounds in the glass was carried out on the basis of X-ray examination.

Name of compounds in solid phase	700°C	900°C	1350°C	1500°C
$NaAlSi_3O_8$ c.s	0.89295	0.86483	–	–
$Al_4B_2O_9$ c.s	0.01188	0.01891	–	–
$Al_6BSi_2O_{13}$ c.s	–	–	0.15674	0.15674

Table 4. Calculation of the thermodynamic stability of glass precursors, option 1 assumption 2

2.1.4. X-ray research of the glasses

The study of X-ray glasses clearly showed amorphous structure, as evidenced increase in the background at low angles and lack of educated, sharp peaks throughout the angular range.

(fig. 4-X-ray glasses of W1). The existence of crystalline barium silicate ($Al_6BSi_2O_{13}$) was not found.

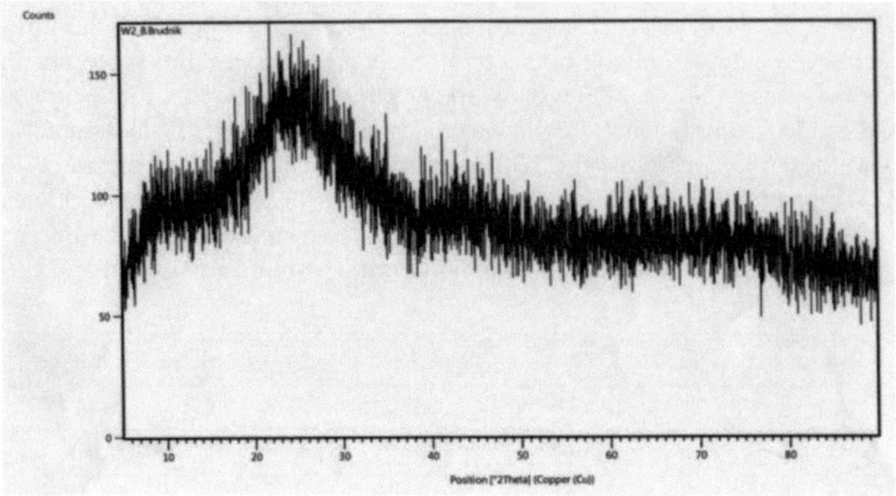

Figure 4. An example of X-ray glass, variant W1

2.1.5. Differential scanning calorimetry research

Thermal analysis of the five types of the glass raw materials and glass of W1 variant was based on differential scanning calorimetry. This method consisted of recording energy required to bring to zero the difference in temperature of the test sample and the reference material as a function of temperature or time.[14] The shape of the DSC curve showed a good agreement with the DTA curve. Endothermic peak was created when the sample temperature was below the standard, and the exothermic peak, when the temperature of the sample was higher than the reference. Samples containing a mixture of raw materials (precursors) of glass heated at a rate of 5° C/min to a temperature of 1400° C. When analyzing the plots, it can be concluded that the incomplete decomposition of the raw materials with the separation of water or CO_2 to a temperature of 350°C occurred, as reflected by the different endothermic peaks. The earliest release of the water of hydration of aluminum nitrate (variant W2-71,0° C, 80,0° C), followed by water coming from the decomposition of boric acid (W1-127,3° C) and CO_2 from the decomposition of sodium carbonate. At a temperature of 548° C on curves of W2 and Ba23 bis inflection appeared to indicate polymorphic transitions of silicon dioxide with the appearance of high gamma phase. For all graphs at about 1230° C and above the melting peak of barium carbonate was observed. Full homogenization mixture of precursors occurred above 1,350° C (W1-DSC chart in fig. 5 and 6)

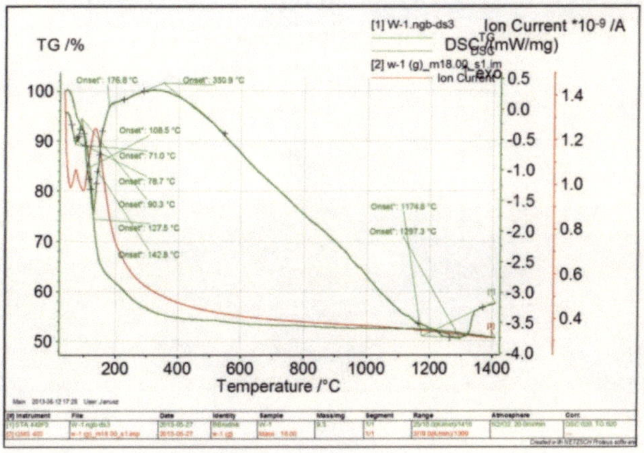

Figure 5. DSC curves of glass raw materials, variant W1

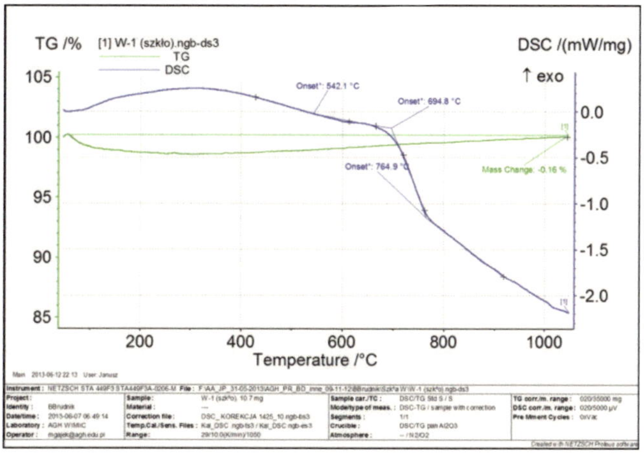

Figure 6. DSC curves of glass, variant W1

Analyzing the DSC plot of W1 glass could be seen of his amorphous. The inflection of the curve reported that there was at 542°C vitrification and 649,8°C started the process of softening. On the thermogravimetric curve TG were not visible weight loss upon heating the glass.

2.1.6. *Wettability research of glasses*

The study of wettability of glasses to substrates of silicon carbide or submicrocrystalline sintered corundum is performed using high-temperature microscope of Leitz- Watzler, type using sessile drop method. Silicon carbide is used as a filler in diamond grinding wheels or supporting grain. SiC substrate, there was degassed under vacuum, and therefore contain adsorbed gases from the air (oxygen, nitrogen), and chronic alcohol bath defatted not entirely surface. Therefore, the observed high temperature sintering (for all variants within 620-670° C glasses) and the temperature drops propagation (920-970° C).

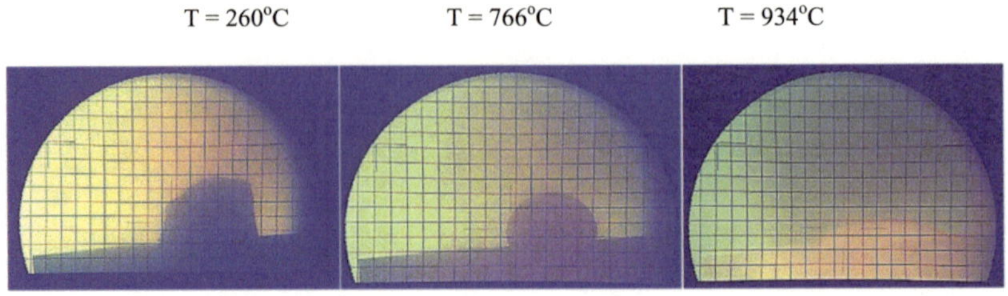

Figure 7. Wetting of cubitron substrate by glass, W1 variant

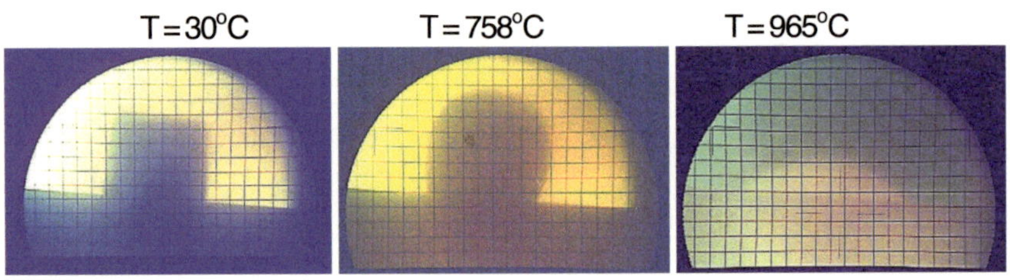

Figure 8. Wetting of silicon carbide substrate by glass, W1 variant

Contact angle theta ranged less than 40 degrees. All the tested glass wetted submicrocrystalline sintered corundum (cubitron) substrate better than the silicon carbide substrate, at lower temperatures (880-930° C), although the same procedure of degreasing were conducted.

2.1.7. Mapping of the surface of glass-substrate microsection

Microscopic observation of transverse cross-sectional views of the glass –SiC substrate or glass-submicrocrystalline sintered corundum substrate carried out using a scanning electron microscope JSM 6460LV and starters EDS (Energy – dispersive X-ray spectroscopy analyzer) revealed the presence of a small transition layer in the system glass- submicrocrystalline sintered corundum (pictures of the glass Ba23 bis-submicrocrystalline sintered corundum) contains barium elements. The phenomena of interlayer (fig. 9) in the case of specimens glass-silicon carbide substrate comprising this was much wider probably resulted from the presence of silicon, both in the glass and substrate.

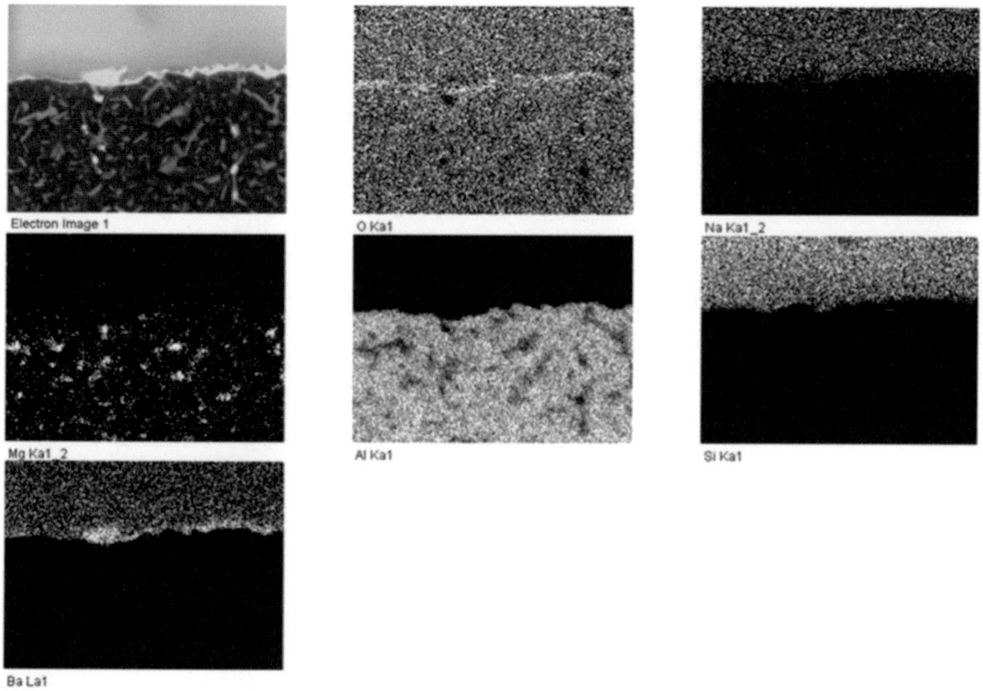

Figure 9. Picture of mapping of the surface of glass-substrate microsection – Ba23 bis variant

2.1.8. Testing the strength of glass

Flexural testing was performed on the four groups of glasses for Zwick Roell Z2.5 machine, at the test speed of 0.5 mm/min, and the spacing supports 34 mm The flexural modulus was determined for all samples by the secant, taking reference points as a force of 50 N and 100 N start the end (Table 5). Maximum bending strength in the range of 100 MPa was obtained for the samples of glass W1 and Ba23 bis. For a group of glasses W4 and W3 it was lower by half. This can be explained by the reduced content of primarily silicon dioxide in these glasses. Similar behavior was observed in the samples of the destruction operation. The largest value of the work of destruction (W1-19,27 N/mm) was obtained for samples W1 and three times lower for samples W3.

Names of samples	Bending strength [MPa] average value	Work of destruction [N/mm] average value	Bending modulus [GPa] average value
W1 glass	105,72 +-2	19.27	25,945
W3 glass	87,08	17,31	14,141
W4 glass	52,29	5,62	21,418
Ba23 bis glass	101,35	16,49	27,195

Table 5. Measurement of flexural strength of glass

2.1.9. *Young modulus measurement of glasses*

The measurements were carried out on the test bench equipped with a flaw, heads broadband and a PC with installed software. Young's modulus and Poisson's ratio of the samples based on the determined velocity of the longitudinal wave and transverse density of the material. All calculations were performed using the program Modulus 1.0. The highest values obtained of glass Ba23 bis (87 GPa) (Table 6). Poisson's ratio for all glasses were the same (0.38).

Name of samples	Poisson's ratio	Young modulus [GPa]	Modulus uncertenainty [%]
W1 glass	0.38	79	2,2
W3 glass	0.38	79	2,2
W4 glass	0.38	76	6,2
Ba 23 bis glass	0.38	87	5,8

Table 6. Measurement results of Young modulus

2.2. Part II Operational tests of grinding wheels containing the newly developed glass

2.2.1. *The starting materials and methods of research*

After summarizing the results of studies on the physical and mechanical properties of the newly developed glass, two of them with the best value properties (Ba23 bis and W1) were selected and used to develop the recipe wheel. Test wheels were prepared to carry out grinding tests. Grinding of flank surface of BNDCC composite samples was performed using wheels 6A2 100x10x4 mm type, containing uncoated diamond grain, Lands LS120 (D46-45-38 microns) type or LS 600F (D25-20-30 microns) with higher concentration of diamond (C125%) and a W1 or Ba23 bis binder (Fig. 9b).

Using DOE, influence on surface roughness and process efficiency of following parameters was derived:

- grinding wheel peripheral speed, v_s = 12, 15, 20 m / s
- working engagement $a_{e:}$ 0.002, 0.005, 0.01 mm / double stroke of the table
- diamond grain D46, D25

Constant parameters during experiments were:

- a research position with the universal tool grinder 3E642,
- the characteristics and dimensions of the grinding type 6A2 100x10x4 mm
- the type of coolant-fed Synkom PGA
- the method by pouring coolant

- maschining time of the one cycle, t = 600 s,
- number of the table double strokes – 3
- table feed speed, v_f = 210 mm / min
- BNDCC composites

The study was carried out on a modernized grinding universal tool grinder 3E642 (fig. 10a), equipped with stepless speed control of the wheel and the cooling system. Grinding process was carried out with cooling by spraying a 2% solution of PGA Synkon coolant concentrate in tap water.

To test of the efficiency of grinding process of custom BNDCC composite sample (boron nitride dispersive in cemented carbide), comprising weight 20% of the grains of cubic boron nitride, with granulation 4-8 mm and dimensions 19x5,7 mm from the edge intersected at a distance of 1 mm from the end, were made at the Warsaw Univeristy of Technology, at the Faculty of Materials Science and Engineering. Before attempts to work the sample was adhered to the brackets in order to attach them to the machine (fig. 10c). Below the picture shows the position of the test, the wheel and bonded composites (fig. 10b).

a) b) c)

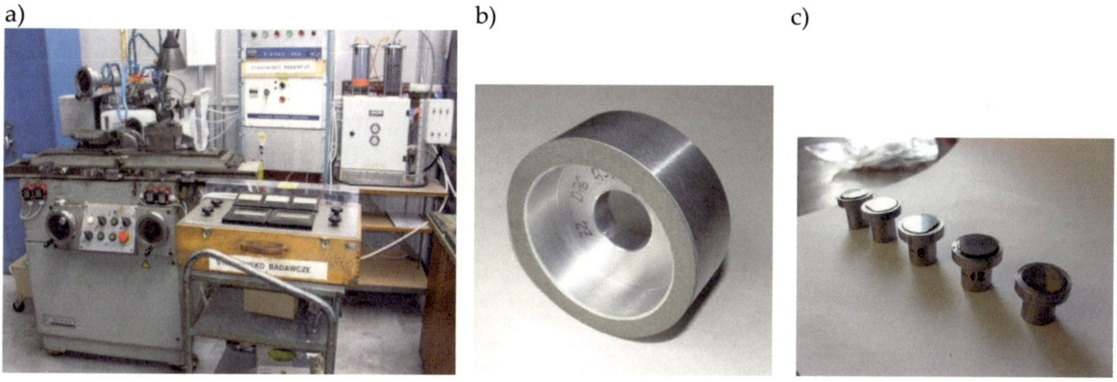

Figure 10. a) The position of a universal tool grinder 3E642 grinding tests with cooling, b) diamond grinding wheel vitrified Ba23 bis c) composites BNDCC intended for research

Measured: the height of abraded material m, height of wheel. Calculated: the volume of abraded material Vw, consumption volume Vs wheel, radial wheel wear Vrs, grinding ratio G, yield losses of (volumetric) Qw, proper performance defects Q'w (deficient performance attributable on the active wheel width).

Sample mass measurements were carried out before and after each treatment using a laboratory scale with an accuracy of measurement 0.001 g. Measurements of the height of wheels and work-before and after each test was performed using electronic calipers accurate to 0.01 mm.

For the measurement of surface topography work-pieces profiler workshop was used. Before and after test were performed via research work on modul profilometer TOPO 01vP designed for measurement and analysis of surface roughness and waviness profiles and the profile of the actual surface without filtration with the program PROFILE. Images of 2D, 3D, marked elevation values R_z, R_t, Ra and horizontal parameters W, St, Sa in accordance with PN-EN ISO 4287th were conducted. The microscopic observations (SEM) of the samples before and after grinding were made. The designated function of the object of research in the form of a polynomial of the second degree of interaction, allowed to determine the statistical relationship between the basic parameters of the treatment and its effects (surface roughness).

2.2.2. Results and discussion

The results of grinding tests of D46 diamond grinding wheels with newly developed binders worked at a peripheral speed of 12, 15, 20 m / s and depth of grinding 0.002, 0.005 mm/double stroke of the table. were presented in Tables 7-8 and figures (fig. 11-16). It was evident that both: the D46 Ba23 bis wheel andW1 D46 wheel were best operating at a depth of 0.005 mm grinding/double stroke. The uniform wear of diamond grains and their self-sharpening were demonstrated.

The grinding wheels with D25 grit size of diamond grains worked at a speed of 15m/s and depth of grinding 0.002 and 0.005 mm/double stroke. All the grinding performance parameters were comparable for both types of wheels. Better results were obtained by D25 grinding wheels with Ba23 bis binder. For the depth of the grinding process the surface roughness of machined BNDCC composites varied within the same limits 0.02-0.03 μm. Comparing the results of performance parameters of the process of grinding wheels with the D46 and D25 granulation can state that they were very similar. Slightly better results were obtained with the wheels with lower diamond grit size (D25).

Kind of wheel	Working engagement a_e, mm	Peripheral speed of wheel vs, m/s	Qw, x10^{-3} mm^3/s	Q'w, x10^{-3} mm^3/mm·s	G [-]	R_a, μm
D46 W1	0,002	12	1,32	0,13	13,56	0,03
D46 Ba23 bis			1,35	0,14	3,41	0,03
D46 W1		15	2,30	0,23	31,12	0,03
D46 Ba23			2,29	0,23	29,48	0,03
D46 W1		20	1,19	0,12	12,02	0,03
D46 Ba23			1,31	0,13	17,91	0,03

Table 7. Results of performance parameters of the process of grinding the surface of composites BNDCC 3,14 D46 grinding wheel W1, Ba23 bis, v_s = 12, 15, 20 m/s with working engagement a_e = 0,002 mm

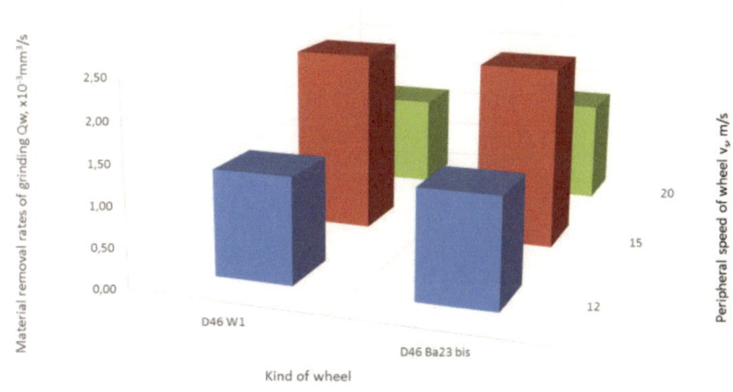

Figure 11. The diagram of material removal rate Qw, x10^{-3} mm^3/ s after grinding of the surface of composites BNDCC 3,14 samples with D46 W1 and Ba23 bis grinding wheels with working engagement a_e: 0.002 mm/double stroke of the table

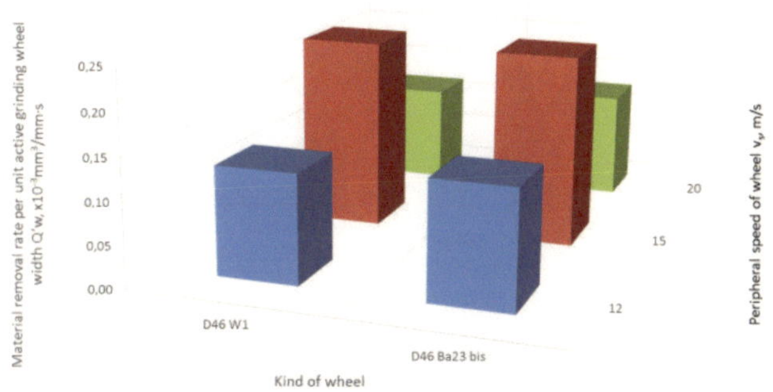

Figure 12. The diagram of material removal rate per unit active grinding wheel width Q'w, x10^{-3}mm^3/mm·s after grinding of the surface of of composites BNDCC 3,14 samples with D46 W1 and Ba23 bis grinding wheels with working engagement a_e: 0.002 mm/double stroke of the table

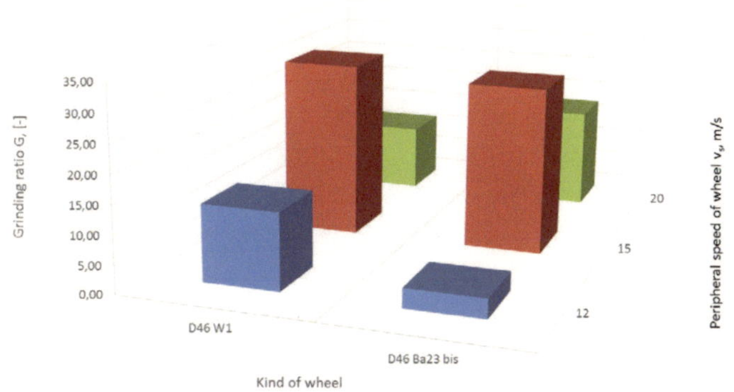

Figure 13. The diagram of grinding ratio G, mm^3/mm^3 for grinding wheels with working engagement ae 0.002 mm/ double stroke of the table

Kind of wheel	Working engagement a_e, mm	Peripheral speed of wheel vs, m/s	Qw, $x10^{-3}$ mm^3/s	Q'w, $x10^{-3}$ $mm^3/mm{\cdot}s$	G	R_a, μm
D46 W1	0,005	12	1,24	0,12	17,39	0,03
D46 Ba23			1,27	0,13	5,88	0,03
D46 W1		15	2,25	0,23	15,24	0,03
D46 Ba23			2,23	0,22	26,94	0,03
D46 W1		20	1,13	0,11	7,82	0,02
D46 Ba23			1,21	0,12	16,94	0,03

Table 8. Results of performance parameters of the process of grinding the surface of composites BNDCC 3,14 D46 grinding wheel W1, Ba23 bis, v_s = 12, 15, 20 m /s with working engagement a_e = 0,005 mm

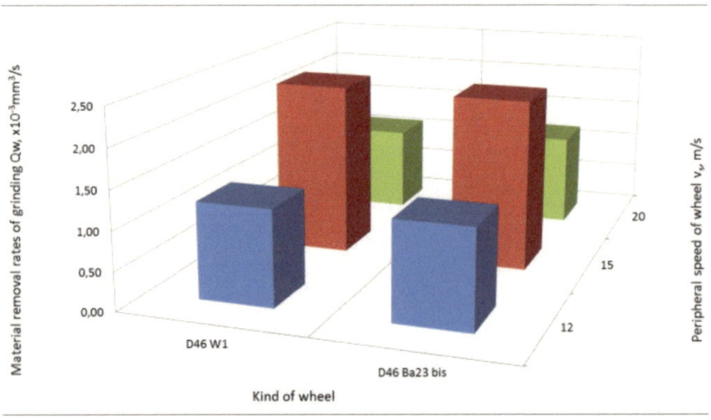

Figure 14. The diagram of material removal rate Qw, $x10^{-3}$ mm^3/s after grinding of the surface of composites BNDCC 3,14 samples with D46 W1 and Ba23 bis grinding wheels with working engagement a_e: 0.005 mm/double stroke of the table

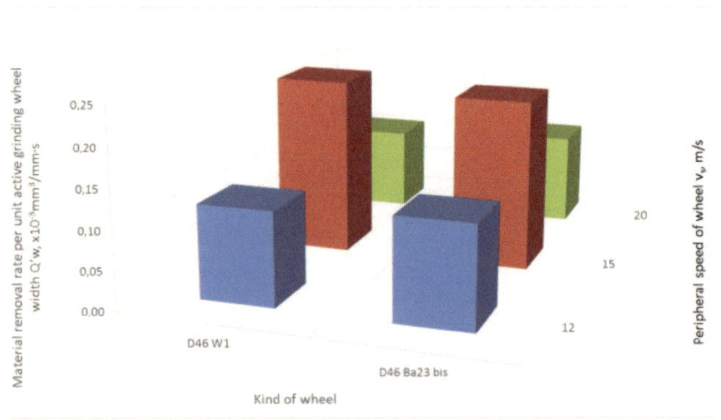

Figure 15. The diagram of material removal rate per unit active grinding wheel width Q'w, $x10^{-3}mm^3/mm{\cdot}s$ after grinding of the surface of composites BNDCC 3,14 samples with D46 W1 and Ba23 bis grinding wheels with working engagement a_e: 0.005 mm/double stroke of the table

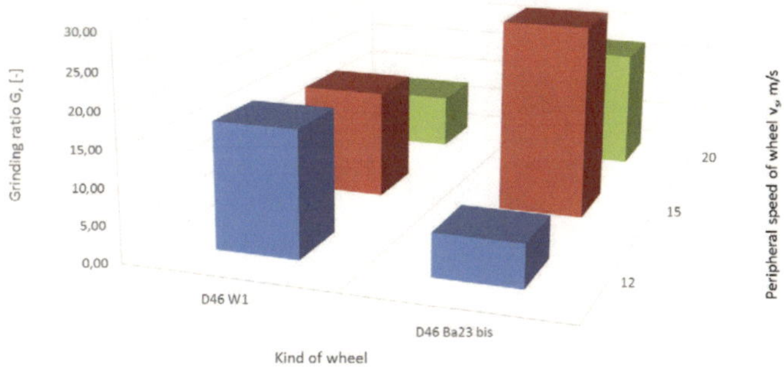

Figure 16. The diagram of grinding ratio G, mm³/mm³ for grinding wheels with working engagement ae 0.005 mm/ double stroke of the table

Kind of wheel	Working engagement a_e, mm	Peripheral speed of wheel vs, m/s	Qw, x10⁻³ mm³/s	Q'w, x10⁻³ mm³/ mm·s	G, [-]
D25 Ba23 bis	0,002	2,31	0,23	31,60	0,02
D25 W1		2,13	0,21	21,91	0,02
D25 Ba23 bis	0,005	2,22	0,22	27,65	0,03
D25 W1		2,02	0,20	27,08	0,02

Table 9. Results of performance parameters of the process of grinding the surface of composites BNDCC 3,13 D25 grinding wheel W1, Ba23 bis, v_s = 15 m/s

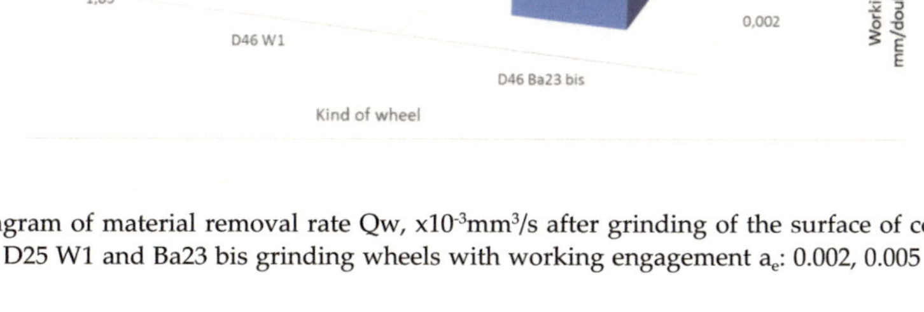

Figure 17. The diagram of material removal rate Qw, x10⁻³mm³/s after grinding of the surface of composites BNDCC 3,13 samples with D25 W1 and Ba23 bis grinding wheels with working engagement a_e: 0.002, 0.005 mm/double stroke of the table

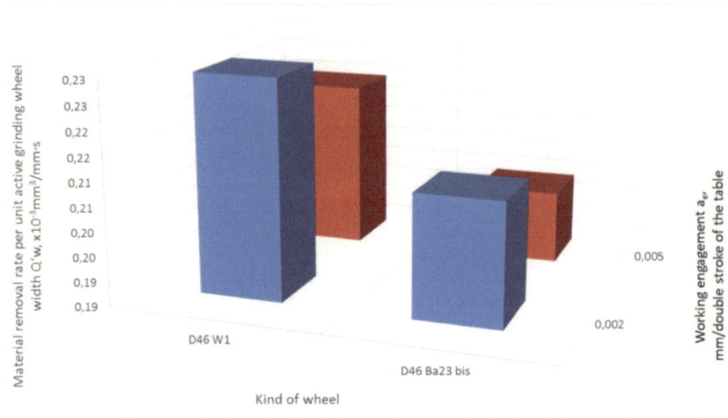

Figure 18. The diagram of material removal rate per unit active grinding wheel width Q'w, x10^{-3} $mm^3/mm \cdot s$ after grinding of the surface of composites BNDCC 3,13 samples with D25 W1 and Ba23 bis grinding wheels with working engagement a_e: 0.002, 0.005 mm/double stroke of the table

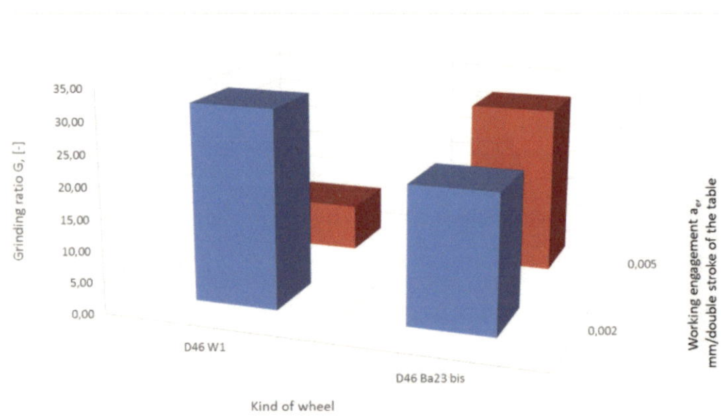

Figure 19. The diagram of grinding ratio G, mm^3/mm^3 for grinding wheels with working engagement ae 0.002 and 0.005 mm/double stroke of the table

2.2.3. Research of geometric structure of BNDCC composites

The study of geometric structure of the samples before and after grinding showed that the assumed test conditions are properly selected (Figure 20a). Machining by grinding the assumed operating parameters of wheel allowed to obtain roughness parameters (R_a, R_z, R_t) times better than the initial (R_a before 1.28-1.40 μm, R_a after-0,018-0,04 μm, R_z before 7,05-7,890 μm, R_z after 0,126-0,150 μm). Image of 3D surface of the BNDCC composite before grinding process showed the presence of significant inequalities (fig. 20a). After grinding process the surface was completely inequalities (fig. 20b).

http://dx.doi.org/10.5772/59503

a)

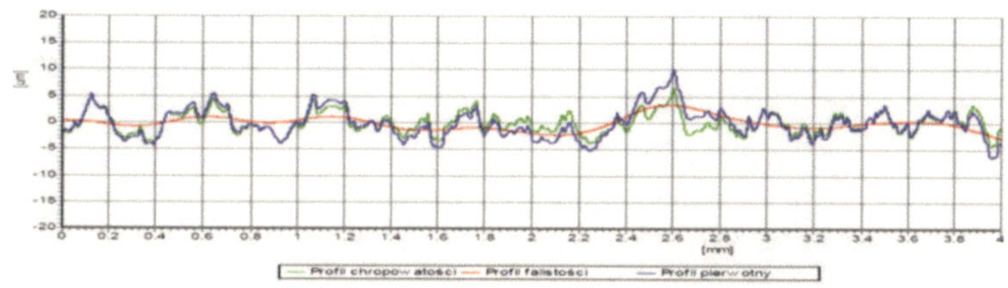

		Średnia	S	Min	Max	Rozstęp
Ra	[µm]	1.248	0.099	1.067	1.655	0.588
Rq	[µm]	1.536	0.112	1.322	1.975	0.652
Rp	[µm]	3.534	0.323	2.903	4.796	1.893
Rv	[µm]	3.556	0.298	2.825	4.220	1.395
Rz	[µm]	7.090	0.517	5.865	8.713	2.848
Rc	[µm]	3.041	0.392	2.482	4.733	2.251
Rt	[µm]	8.811	0.859	6.795	11.061	4.266

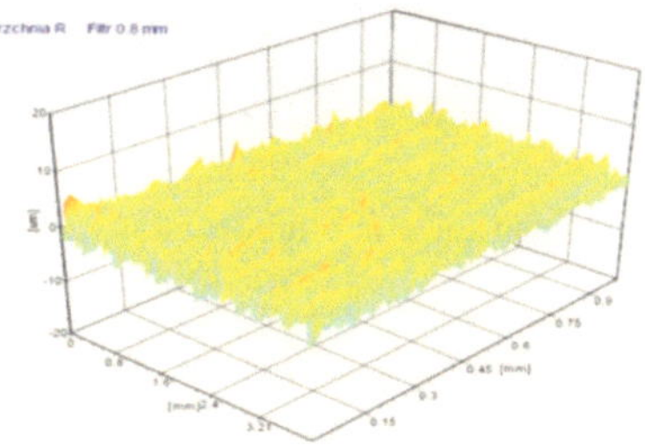

b)

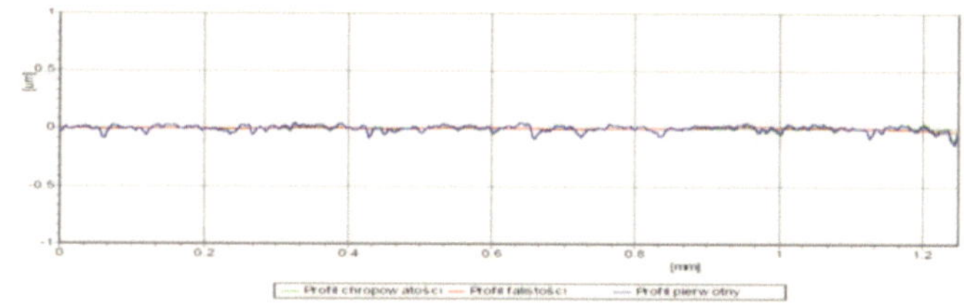

		Średnia	S	Min	Max	Rozstęp
Ra	[µm]	0.018	0.001	0.015	0.022	0.006
Rq	[µm]	0.024	0.002	0.020	0.029	0.009
Rp	[µm]	0.046	0.005	0.037	0.062	0.025
Rv	[µm]	0.080	0.011	0.060	0.122	0.062
Rz	[µm]	0.126	0.014	0.100	0.173	0.073
Rc	[µm]	0.051	0.006	0.039	0.067	0.027
Rt	[µm]	0.171	0.031	0.117	0.247	0.129

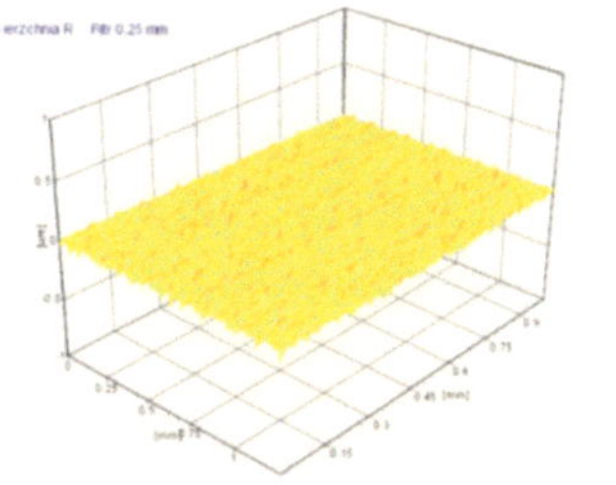

Figure 20. The results of the surface roughness of the composite BNDCC polished 1.47 grinding D25 Ba23: a) before grinding, b) after grinding

2.2.4. The microscopic image of BNDCC composite

The samples of BNDCC composites were observed under a scanning microscope before and after the grinding tests. For all composite samples before grinding process parallel scratches were visible on the surface of the samples. These were probably established through their mechanical treatment after sintering. Microscope image of the samples after grinding showed sharply defined edges with blunted of blade grains in carbide matrix.

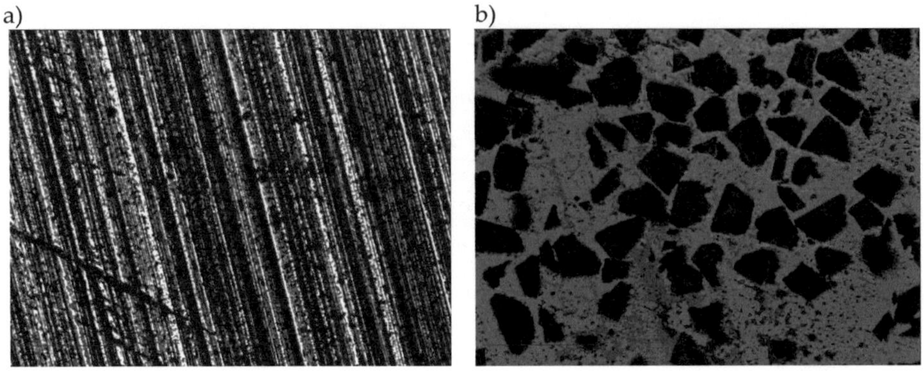

Figure 21. Microscope image of composite BNDCC: a) before; b) after the treatment grinding D46 Ba23 bis binder

2.2.5. The mathematical model of the function of research object

The mathematical model describing the change in the surface roughness Ra of the treated object as a function of the depth of grinding and peripheral speed of grinding wheel was shown in graphic form (fig. 22, 23). The function of the object of research was determined for the grinding process of D46 Ba23 bis wheel and D25W1 wheel BNDCC composite of grit size 4-8 m assuming independent parameters (depth of grinding, ae and grinding wheel peripheral speed vs) and the dependent parameter (surface roughness of the composite). For DOE purposes and analysis of experimental results, STATISTICA was used [15,16]. The range of variability was low. Statistical analysis of the results of experimental studies included

- Approximation of object function tests,
- Statistical verification of the adequacy of the function approximating,
- Statistical verification of significance approximating function coefficients.

The function of the object of research adopted in the form of a polynomial of the second degree of interaction is described by the following formula:

for D25 Ba23 bis

$$y = 0,15 - 7,76\, a_e - 0,01\, V_S + 0,27\, a_e V_S$$

for D25 W1:

$$y = 0,014 - 1,6\, a_e - 0,02\, V_S + 0,18\, a_e V_S \backslash$$

where:

a_e, vs – the size of the input,

$a_0 \div a_5$– polynomial coefficients.

The results of preliminary calculations showed that the best fit regression equations to the results of the experiment allows the second degree polynomial model of interaction, and therefore the calculated regression equation stepwise regression presented in this form. Analysis of individual regression equations fit to the experimental results was based on multivariate correlation coefficient R and also based on function values and Student's t-value of F-Snedecor. The level of significance p = 0.05.

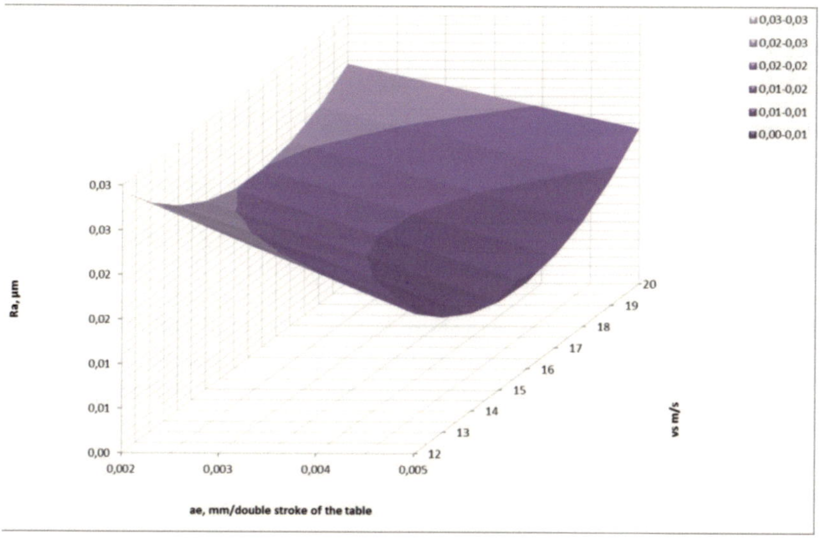

Figure 22. The influence of processing parameters on the surface roughness of the composite samples BNDCC polished by grinding wheel D46 with Ba23 bis

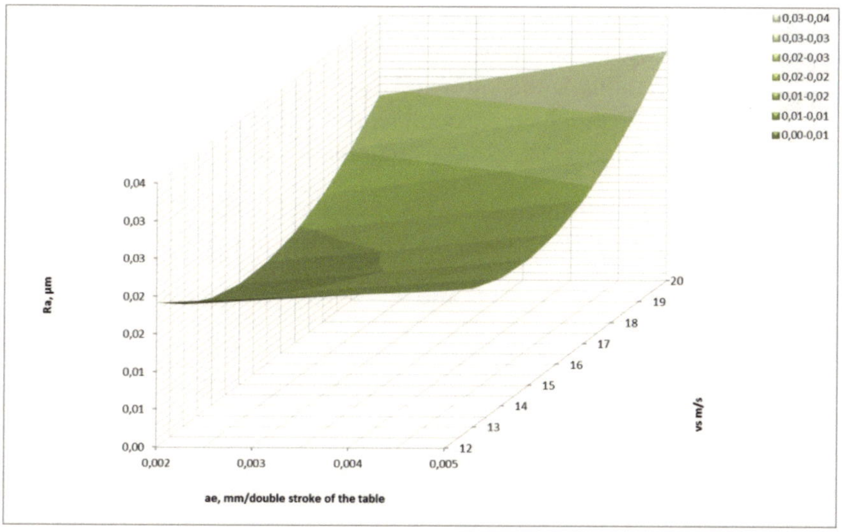

Figure 23. The influence of processing parameters on the surface roughness of the composite samples BNDCC polished by grinding wheel D46 with W1 binder

Analysis of relationship shown in Figure 22 and 23 confirmed that the minimum value of surface roughness BNDCC samples was obtained at the peripheral speed of grinding wheels in the range 14-16 m/s and 0.005 mm depth of grinding/double stroke of table. These data verified the results of performance tests of these samples.

3. Summary and conclusion

Based on the survey it can be concluded that:

- the research glass fulfilled their utility criteria as binders for super hard abrasive tools;
- glass received by frytting method belonged to the group of light glasses (ϱ <2.64 g / cm 3). It was completely transparent with a bluish tint;
- thermodynamic calculation of chemical stability of glass precursors were conducted by VCS algorithm. It was shown in accordance with first assumption that the three compounds were stable in solid states at high temperatures: barium silicate, silicon dioxide, aluminum borosilicate; in the second assumption only one (barium borosilicate). X-ray crystallographic studies have not confirmed the presence of compounds. Investigated samples were in pure amorphous form;
- differential scanning calorimetry of precursor glasses showed their distribution with the separation of water or CO_2 to a temperature of 350°C. The whole sets underwent complete homogenization above 1,350° C;
- substrates were wetted well by all variants of glasses, although in all cases submicrocrystalline sintered corundum was wetted better than silicon carbide. Microscopic observation of samples transverse glass-substrate systems contains submicrocrystalline sintered corundum after the wettability studies showed the presence of narrow interlayer contained barium. In -case of transverse specimens with silicon carbide substrates were much wider, probably resulted from the presence of -silicon in glass and substrate;
- the flexural strength showed the highest values was obtained by W1 and Ba23 bis glass. The work of destruction was tree times lower for W3 glass than W1 glass;
- diamond wheels with D46 or D25 grains and Ba23 bis or W1 binders fulfilled the criteria of utility in the flat grinding process;
- analysis of grinding process efficiency Q_w [$x10^{-3}mm^3/s$] Q'_w [$x10^{-3}mm^3/mm \cdot s$], G (listed in Tables 3, 4) showed a very good result under the processing conditions. It was observed that the most favorable results are obtained when the samples of BNDCC composite was machining in – the grinding depth of 0.005 mm/double stroke of table and the peripheral speed of the wheel of 15 m/s;
- the results of the geometrical structure of the surface of the BNDCC composites showed that the grinding surface roughness was R_a 0.018 -0.04 μm compared to the initial roughness of

http://dx.doi.org/10.5772/59503

R_a 1.248 -1.4 μm, R_z after grinding 0.12-136 μm compared to the initial value of R_z 7,096-7,96 μm;

- the mathematical model of grinding process properly describes the working condition necessary to obtain the lowest roughness.

4. Conclusions

- the use of diamond wheels with newly developed vitrified (ceramic) binder in conventional grinding technology has allowed to obtain a mirror of BNDCC composites.

Presented research results were obtained within the project: "Application of modern BNDCC and DDCC composites for cutting tools", Applied Research Program – Contract No. PBS1/A5/7/2012 financed by The National Centre for Research and Development in Poland

Author details

Barbara Staniewicz-Brudnik*, Elżbieta Bączek and Grzegorz Skrabalak

*Address all correspondence to: barbara.brudnik@ios.krakow.pl

Institute of Advanced Manufacturing Technology -011 Kraków, Wroclawska, Poland

References

[1] Kruszewski M, Rosinski M., Grzonka J., Ciupiński Ł., Michalski A., Kurzydłowski K J: Cu composites - a diamond with a high thermal conductivity produced by PPS, Ceramic Materials, 64, 2012, 3, 335-337

[2] Lin K.H, Peng SF, Lin S.T Sintering parameters and wear performances of vitrified bond diamond grinding wheels International Journal of Refractory Metals and Hard Materials, [W] Elsevier, vol 25, 2007, 25-31.

[3] Rosiński M, Chrzanowski W,. Spychalski M, Michalski A. Influence of process parameters on the grinding surface of the composite state WCCo / diamond produced by PPS Ceramic Materials, 64, 2012, 3, 314-318

[4] Rosiński M., Michalski A.: WCCo / cBN composites produced by pulse plasma sintering method, Journal of Materials Science, vol. 47, no 20, pp. 7064-7071, 2012, DOI 10.1007 / s10853-012-x

[5] Rosiński M, Wachowicz J., Ściegienko A., Michalski A. The new composite material is a diamond in the matrix of tungsten carbide for cutting tools for working wood, Ceramic Materials, 64, 2012, 3, 329-332

[6] Staniewicz-Brudnik B.,. Plichta J, Nadolny K., Pluta J.: The study of CBN grinding wheels grinding efficiency with increased porosity of micrograins of spherical corundum. Archive of Production Engineering and Automation, 2006, Vol. 26, No. 2, p. 77-84. Poznan: Poznan University of Technology Publishing House, 2006.

[7] Staniewicz-Brudnik B., Plichta J, Nadolny K.: Effect of porous glass-ceramic materials Addition on the cubic boron nitride (cBN) tools properties. Optica Applicata, 2005, Vol. XXXV, No. 4, p. 809-817., 2005.

[8] Staniewicz-Brudnik B.: grinding with cubic boron nitride ceramic bond. IZTW Bulletin, 2005, No. 1, p. 31-34. Krakow: IZTW, 2006.

[9] Wang Y. et al. Influence of zinc particles on oxidation resistance of diamond / borosilicate glass compositions Diffusion and Defect Data PTB: Solid State Phenomena 175, 2011, 8-12.

[10] Bain Report The Global Diamond Industry

[11] Houy G.: Effect of porosity on the grinding performance of vitrified bond diamond grinding wheels for PCD blades Ceramics International 2012, http: // dx doi.org/ 101016/ceramint.2012304.074

[12] Procyk B, Staniewicz-Brudnik B., Majewska-Albin K.: Investigations of wettability and reactivity in glass / carbon and glass / ceramic systems. [W:] HTC-2000 High Temperature Capillarity, Third International Conference, 19-22 November, Kurashiki, Japan. Transactions of JWRI, 30, 2001, 149-154. [W]: Osaka University: 2001

[13] Jackson M. J., Hitchiner M. P. High Performance Grinding and Advanced Cutting Tools chapter 2 [W]: Springer Science + Business Media New York, 2013.

[14] Partyka J. Sitarz M., Leśniak M., Gasek K., Jeleń P.: The effect of SiO_2/Al_2O_3 ratio on the structure and microstructure of the glasses from SiO_2–Al_2O_3–CaO–MgO–Na_2O–K_2O system Spectrochimica Acta Part A: Molecular and Biomolecular Spectroscopy Volume 134, 5 January 2015, pp.621–630, DOI: 10.1016/j.saa.2014.06.06

[15] Polanski Z.: Optimization methods in machine technology, PWN, Warsaw, 1977

[16] PN – EN 4287

Chapter 5

An Overview of Densification, Microstructure and Mechanical Property of Additively Manufactured Ti-6Al-4V — Comparison among Selective Laser Melting, Electron Beam Melting, Laser Metal Deposition and Selective Laser Sintering, and with Conventional Powder Metallurgy

Ming Yan and Peng Yu

Additional information is available at the end of the chapter

http://dx.doi.org/10.5772/59275

1. Introduction

Additive manufacturing (AM) is the newest powder metallurgy (PM) technique [1-3] while Ti-6Al-4V is the single most important Ti alloy [4-9]. AM Ti-6Al-4V is being widely pursed by research community and industry for its capability to produce complicated, net-shape engineering parts and/or customised, biomedical implants.

This chapter begins with an introduction of the fundamental properties of Ti-6Al-4V, and its densification mechanism, typical microstructure and mechanical property achievable by conventional PM routes. This functions as a point of reference for the following discussion of the AM Ti-6Al-4V in terms of densification, microstructure, and mechanical property. The mostly popular laser-based AM techniques, namely selective laser melting (SLM), electron beam melting (EBM), laser metal deposition (LMD) and selective laser sintering (SLS), for the fabrication of Ti-6Al-4V have been overviewed based on an analysis of over 100 individual studies. Heat treatment is essential to most of the AM Ti-6Al-4V. Principles for selecting appropriate heat treatment for the AM Ti-6Al-4V are proposed based on martensite phase transformation and optimisation of mechanical properties. Oxygen impurity is an issue to most Ti materials and it is addressed in this chapter as well; counter measurements to mitigate oxygen have been suggested which involves the use of rare earth based materials.

2. Preliminary knowledge: Fundamental properties of Ti-6Al-4V and conventional PM Ti-6Al-4V

2.1. Densification of PM Ti-6Al-4V and typical microstructure and mechanical property

- **Sintering mechanism:** Sintering/densification of conventional PM Ti-6Al-4V is mostly through solid state sintering. Sintering temperatures are normally selected in the β phase region [10-13], at a temperature (e.g. at 1300ºC) well below the liquidus temperature (~1660 ºC for Ti-6Al-4V, Table 1) [14,15]. For sintering under pressure such as via hot pressing or spark plasma sintering, the sintering temperature can be lower, for instance, around 900°C [16]. Driving force for densification during solid state sintering is reduction of surface area and surface free energy by eliminating solid-vapour interfaces [17-19]. Sintering procedure mainly involves diffusional flow of composing elements, including surface diffusion, grain boundary diffusion and lattice diffusion [17-19].

- **Sintering activation energy:** Using the master sintering curve approach, Crosby [20] estimated that sintering activation energy, the Q value, of Ti-6Al-4V is about 130 kJ/mole. This is consistent with the reported Q value of the self-diffusion of titanium (92.5 kJ/mole-158 kJ/mole) over the β-Ti range from 900 to 1250 ºC, affirming that the densification of Ti-6Al-4V is mainly controlled by self-diffusion of titanium [8].

- **Sintered density:** Most of the as-sintered Ti-6Al-4V materials show 95%-99% of theoretical density [4-13]. The as-sintered density depends on a few factors such as compaction pressure and powder size. Using TiH_2 as the starting powder may assist in improving the as-sintered density [21,22], but to achieve a pore-free alloy (no less than 99.8% of theoretical density) post treatment such as via extrusion or hot isostatic pressing (HIP) is necessary [23].

- **Microstructure of PM Ti-6Al-4V**: PM Ti-6Al-4V is expected to show the following microstructural characteristics [24]: (a) The overall microstructure is close to the equilibrium state due to low cooling rate adopted for most sintering practices. (b) There could be some annealing and/or aging effect resulting from the slow cooling process from the isothermal sintering temperature to room temperature. Aging-induced phases such as isothermal ω may form during cooling. (c) Pores are part of the as-sintered microstructure due to the difficulty to achieve a pore-free microstructure in most cases. Fig. 1(a) provides a typical SEM image of the as-sintered PM Ti-6Al-4V, consisting of grain boundary α, α lath and β phases [25]; the overall volume fraction of the α phase is more than 85% [10-13].

- **Mechanical property of PM Ti-6Al-4V:** Mechanical properties of the PM Ti-6Al-4V are highly dependent upon oxygen level. Typical cases are shown in Figure 1(b) to demonstrate ductility of PM Ti-6Al-4V as a function of oxygen [26]. The fracture strength of PM Ti-6Al-4V is comparable or even higher than the wrought material (ASTM B348) [9].

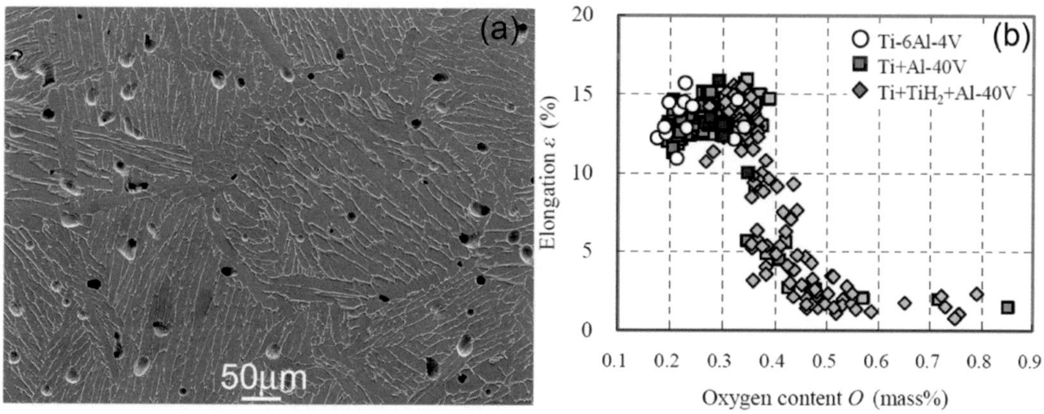

Figure 1. (a) Scanning electron microscopy (SEM) image to show the typical microstructure of PM Ti-6Al-4V [25] and (b) ductility of PM Ti-6Al-4V as a function of oxygen [26].

2.2. Ti-6Al-4V: Basic physical and mechanical properties

Table 1 and Table 2 list the fundamental physical and mechanical properties of the Ti-6Al-4V, respectively [14,15,27,28]. They serve as a point of reference and provide benchmark values for the following discussion on the AM Ti-6Al-4V.

Physical property	Value
Density of solid (ρ)	4.43 g/cm^3
Density of liquid (ρ)	3.89 g/cm^3
Solidus temperature	1877 K (1604°C)
Liqudus temperature	1933 K (1660°C)
Temperature of (α+β) → β	1253K (980°C)
Temperature of α → β	~1023K (750°C)
Thermal conductivity of solid (k_s)	6.7 W/m/K
Thermal conductivity of liquid (k_l)	32.5 W/m/K
Specific heat capacity of solid (Cp_s)	0.526 J/g/K
Specific heat capacity of liquid (Cp_l)	0.872 J/g/K
Coefficient of thermal expansion of solid (K^{-1})	8.6 μm/m/K
Temperature of martensite phase transformation (M_s)	~1053K (780°C) or ~883K(610°C)

Table 1. Fundamental physical properties of Ti-6Al-4V [14,15,27,28]

Mechanical property	Value
Tensile strength, yield ($\sigma_{0.2}$)	880 MPa
Tensile strength, Ultimate (UTS)	950 MPa
Elongation (ε)	14 %
Reduction of Area (R)	36%
Hardness, Hv	349
Young's modulus (E)	113.8 GPa
Poisson's ratio (υ)	0.342
Fatigue strength (at $1*10^7$ cycles and $K = 3.3$)	240 MPa
Fatigue strength (unnotched at $1*10^7$ cycles)	510 MPa
Fracture toughness	75 MPa•m$^{1/2}$

Table 2. Mechanical properties of Ti-6Al-4V achievable via forged- then-annealed [14,15,27,28]

3. Various AM approaches for Ti-6Al-4V and their key processing parameters

3.1. Selective Laser Melting (SLM)

There are a variety of suppliers providing reliable SLM machines for producing AM Ti-6Al-4V [29-46]. EOSINT M270, Trumpf LF250, and SLM Solutions are well known SLM facilities. Figure 2 shows a schematic graph to present the working principle of the SLM [46]. Table 3 provides a summary about the key machine parameters and working conditions via a typical SLM machine, SLM 280 HL [29-46].

	Parameter
Typical equipment	SLM 280HL
Build volume	280 mm * 280 mm * 350 mm
Laser type	YLR Fibre Laser
Laser beam size	70 μm - 120 μm
Atmosphere during processing	With protection by Ar, N_2 or He
Scanning speed	Up to 15000 mm/s (mostly 125 mm/s – 802 mm/s)
Feed type	5 μm – 50 μm
Layer thickness	Mostly 20 μm – 75 μm
Build temperature	Mostly naturally heated
Build density	99.7% - 100%
Substrate	Without substrate; or with self-substrate heated up to 700°C

Table 3. Typical technical parameters of the SLM [29-46].

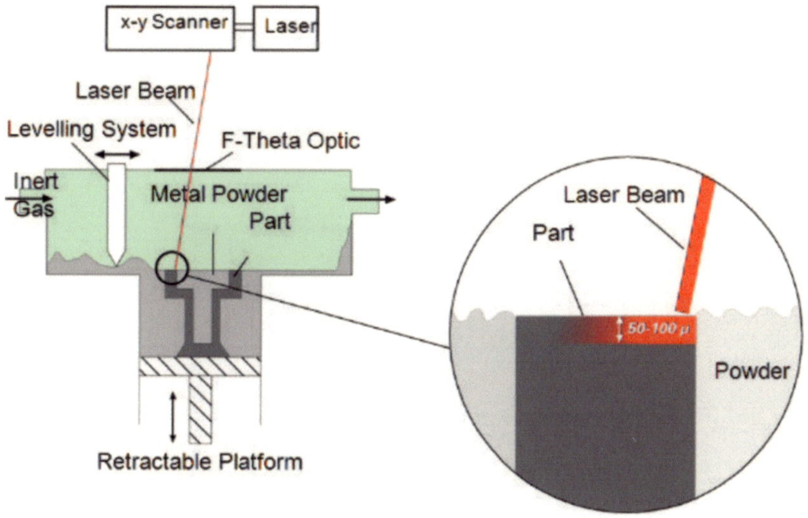

Figure 2. Schematic graph to show the working principle of the SLM [46].

3.2. Electron Beam Melting (EBM)

Arcam provides a variety of EBM machines for processing Ti and Ti alloys, including Arcam A1, A2, S12 and S400, and it dominates the market for the time being [47-66]. Figure 3 provides a schematic graph to show the working principle of the EBM [66]; key machine parameters and working conditions are summarised into Table 4 via a typical example of Arcam A2 EBM.

	Parameter
Typical equipment	Arcam A2
Build volume	250 mm * 250 mm * 400 mm or Φ300 mm * 200 mm
Powder supply	7 kW max.
Laser beam size	200 μm – 1000 μm (mostly 100 μm - 500 μm)
Atmosphere during processing	10^{-1} Pa - 10^{-3} Pa, or with protection by partial pressure of He
Scanning speed	Up to 8000 mm/s (mostly 125 mm/s – 802 mm/s)
Feed type	25 μm – 149 μm (40μm - 100μm standard)
Layer thickness	20 μm – 100 μm
Build temperature	640°C - 700°C
Build density	99.4% - 100%
Substrate	Stainless steel, normally heated to 720°C - 800°C

Table 4. Typical technical parameters of the EBM [47-66].

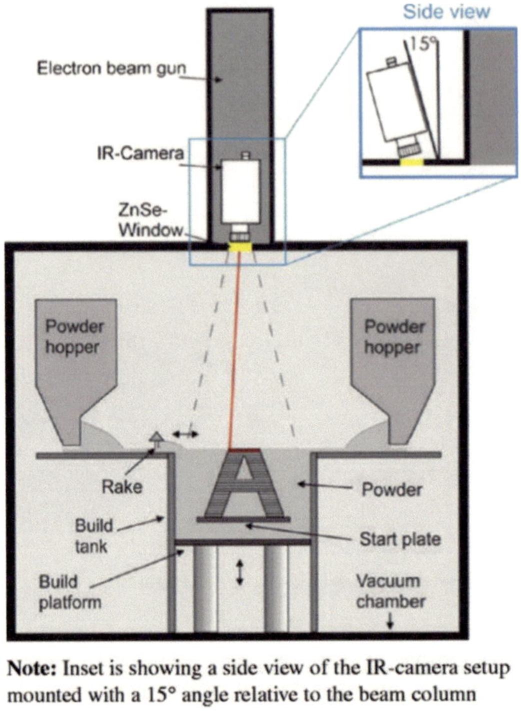

Figure 3. Schematic graph to show the working principle of an EBM equipment [66].

3.3. Laser Metal Deposition (LMD)

A variety of LMD machines and laser systems are being used for the fabrication of metals and alloys including Ti-6Al-4V, such as LEMS TM Nd:YAG, Trumpf HLD 3504 Yb:YAG and IPG with Yb fibre laser [67-78]. Figure 4 and Table 5 present the general working principle, working conditions and key machine/processing parameters.

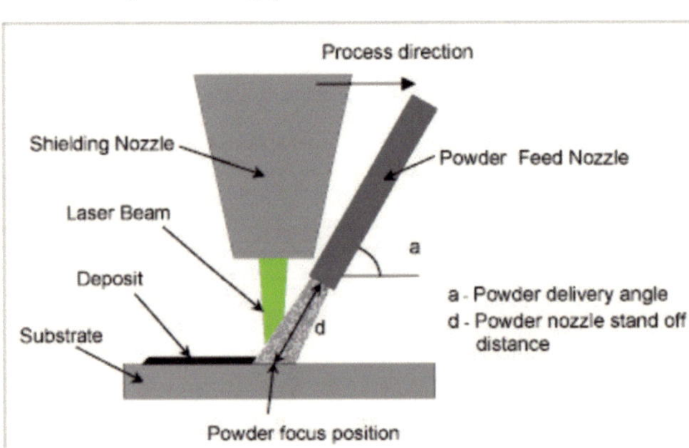

Figure 4. Schematic graph to show the working principle of an LMD equipment [76].

	Parameter
Typical equipment	Trumpf HLD 3504
Build volume	250 mm* 250 mm * 400 mm
Powder supply	3.5 kW max.
Laser beam size	Mostly 500 μm - 4100 μm
Atmosphere during processing	Argon flow protection
Scanning speed	Mostly 2 mm/s – 40 mm/s
Feed type	Mostly 10 μm – 200 μm (powder); 1.2 mm - 1.6 mm wire
Layer thickness	40 μm – 100 μm
Build temperature	Naturally heated
Build density	Can be >99.9%
Substrate	Ti-6Al-4V or Ti; room temperature or preheated to 200°C

Table 5. Typical technical parameters of the LMD [67-76].

3.4. Selective Laser Sintering (SLS)

EOSINT M250X and EOISNT M270 are being used as the SLS machines for preparing AM Ti-6Al-4V [79-88]. The general working conditions and key machine/processing parameters are shown by Figure 5 and listed in Table 6.

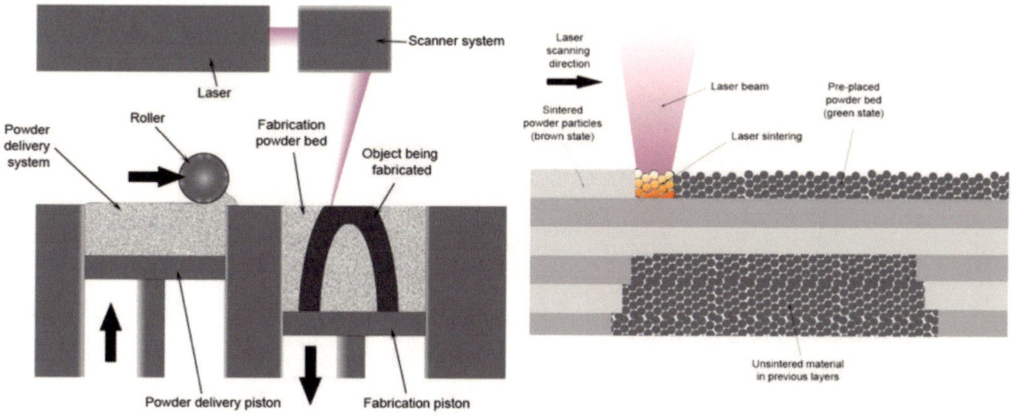

Figure 5. Schematic graph to show the working principle of an SLS equipment [88].

	Parameter
Typical equipment	EOSINT M270
Build volume	250 mm * 250 mm * 215 mm
Laser type	Yb-fiber laser, 200 W
Powder supply	5.5 kW max.
Laser beam size	100 μm - 500 μm
Scanning speed	Up to 7000 mm/s (mostly 50 mm/s - 100 mm/s)
Layer thickness	20 μm - 100 μm
Feed type	Mostly 37 μm - 74 μm (should be similar to SLM)
Build temperature	- (can use preheated powder at e.g. 600°C)
Build density	Mostly below 99%
Substrate	Normally Ti, can be heated to e.g. 230 °C

Table 6. Typical technical parameters of the SLS [77-88].

4. Densification of AM Ti-6Al-4V

4.1. As-built density of AM Ti-6Al-4V

Figure 6 summarises reports on as-built density of the AM Ti-6Al-4V using LMD, SLM, EBM, and SLS [29-88]. The figure shows that the as-built density of AM Ti-6Al-4V is mostly higher than 99% of theoretical density and the material can be fully dense if processing parameters are appropriately selected. SLS Ti-6Al-4V is exceptional to this conclusion, which as-built densities are around 95% and similar to the as-sintered density of the conventional PM Ti-6Al-4V [4-13].

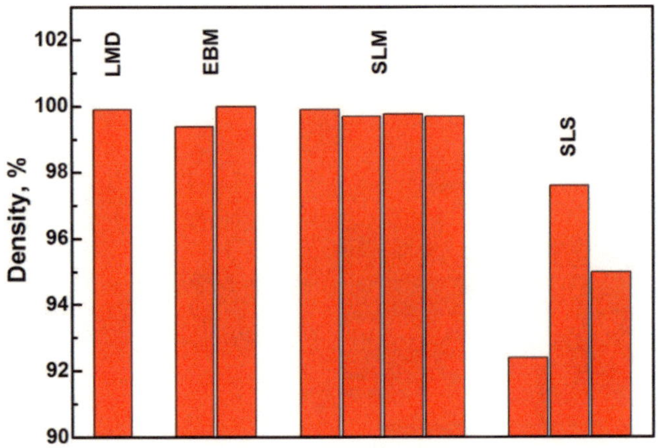

Figure 6. Density of as-built AM Ti-6Al-4V [29-88].

4.2. Densification of SLS Ti-6Al-4V: Solid state sintering and liquid phase sintering

The fact that the as-built density of the SLS Ti-6Al-4V is similar to that of the conventional PM Ti-6Al-4V (see Figure 6) implies the two processing approaches should have analogous densification mechanism. This is understandable by referring to Figure 5 which shows that the SLS processing uses laser as the heating source while for the conventional PM Ti-6Al-4V, heating is mainly through conductive and/or radiation heat. Other than this the two processing pathways are essentially same. This further suggests that the driving force and activation energy for densification of the conventional PM Ti-6Al-4V should be applicable to the SLS Ti-6Al-4V. During SLS of Ti-6Al-4V, however, there is possibility that liquid phase sintering (LPS) occurs due to local overheating to temperatures higher than the liquidus temperature of Ti-6Al-4V (~1660°C). LPS provides extra driving force for sintering for it has extra surface energies, i.e. resulted from liquid surface and from liquid-solid interface [89,90]. It involves different sintering mechanisms, e.g. pore-filling, from the solid state sintering. These normally can contribute to an as-sintered density of up to 99.5% high [89,90]. Since the as-sintered density of the SLS Ti-6Al-4V is lower than this value, the overwhelming densification mechanism for the SLS Ti-6Al-4V should still remain as that of the solid state sintering.

4.3. Densification of EBM, SLM, and LMD Ti-6Al-4V: Solidification from liquid

Full or nearly full denseness is achievable in EBM, SLM and LMD Ti-6Al-4V (see Figure 6), implying that the densification mechanism for these AM approaches is different from that of

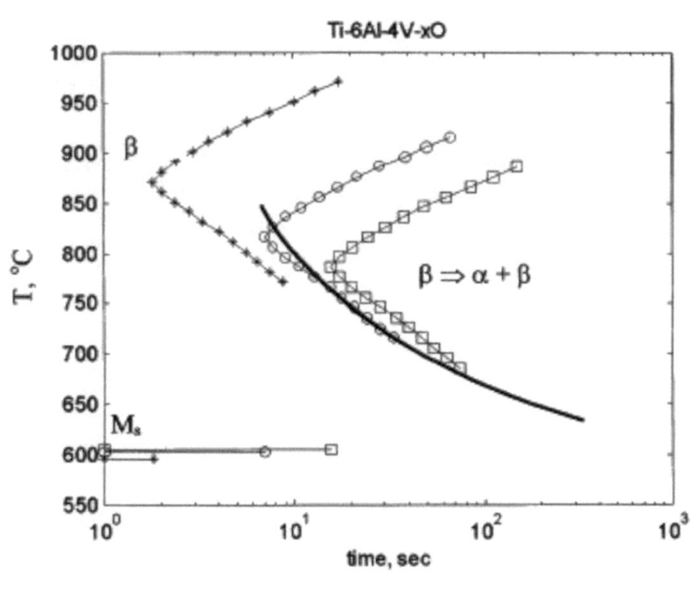

Figure 7. Simulated T-T-T curve of Ti-6Al-4V, which can be used to forecast the phase selection during solidification of the alloy [91]. M_s in the figure denotes the martensite phase transformation. Open squares in the figure are for 0.05wt. %O, circles for 0.1 wt.% and stars for 0.2 wt.%O. The solid line is drafted based on experimental results. Oxygen (O) is found to be able to lower the temperature for the martensite phase transformation.

the SLS. Indeed, for these three laser-based processing approaches, the densification process is more of solidification from liquid rather than the normal sense of sintering. The so-called time-temperature-transformation (T-T-T) of Ti-6Al-4V regulates phase selection and phase constitution of the solidified microstructure. Figure 7 provides the T-T-T curve of the Ti-6Al-4V alloy containing different levels of oxygen [91]. Under equilibrium conditions (i.e. low cooling rates), the resultant microstructure will be a mixture of thermodynamically stable α and β phases, while high cooling rates can enable formation of martensite phases.

5. Microstructure of AM Ti-6Al-4V (As built)

5.1. Microstructure of SLS Ti-6Al-4V

SLS is largely similar to the conventional PM and this explains that the microstructure of SLS Ti-6Al-4V is close to that of equilibrium state as with the PM or cast Ti-6Al-4V [79-88]. Figure 8 compares the microstructure of SLS Ti-6Al-4V with that of an HIP Ti-6Al-4V [81]. In general both materials are of lamellar structure consisting of (α+β) phases and show no evident difference to each other.

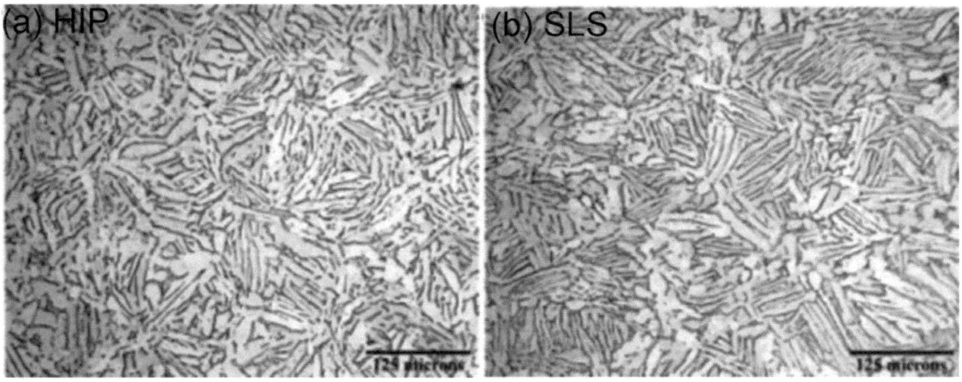

Figure 8. Optical microscopy images of (a) an HIP Ti-6Al-4V and (b) an SLS then HIP Ti-6Al-4V. The latter generally presents a lamellar type of microstructure and almost identical to the former [81].

5.2. Microstructure of EBM, SLM and SMD Ti-6Al-4V

5.2.1. Solidification map

As aforementioned, regarding EBM, SLM and SMD, the AM processing is essentially a solidification process [29-78]. In this regard, the solidification map (see Fig. 9) suggested by Kobryn and Semiatin [1] is useful for predicting the solidification microstructure and can even be served for microstructural design. The two parameters, *R* and *G*, that are crucial for composing the solidification map can be measured and/or calculated as follows:

$$R = \mathrm{d}z / \mathrm{d}t \tag{1}$$

$$G = \mathrm{d}T / \mathrm{d}z \tag{2}$$

where R is the solidification velocity which can be calculated based on a distance (dz) moved by solidus isotherm over a certain amount of time (dt), and G is the thermal gradient which can be obtained using a temperature window (dT) between solidus and liquids isotherms over a certain amount of distance (dz).

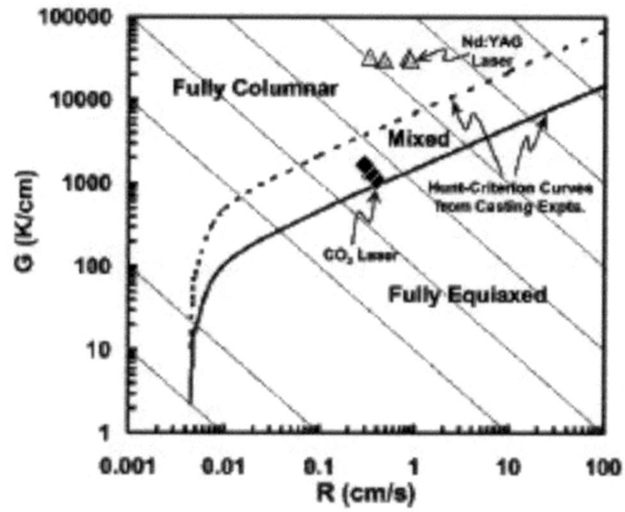

Figure 9. Solidification map of Ti-6Al-4V with simulated laser-glaze data points [1].

5.2.2. Texture and heterogeneity issue

Texture and heterogeneity are widely observable microstructural features in EBM, SLM and SMD Ti-6Al-4V [29-76]. They are formed mainly due to the following two reasons: (a) different temperature distribution in the as-built sample and (b) different cooling rates in the various parts of the sample.

Figure 10 provides optical graphs for an EBM Ti-6Al-4V at the transverse direction (Fig. 10a) and longitudinal direction (Fig. 10b) [4]. The former shows a lamellar microstructure while the later presents directional growth along the build direction. Greater thermal gradient in the latter is suggested to be the main reason for the formation of this columnar type of microstructure [4]. Figure 11 shows other two examples to illustrate the microstructural heterogeneity between the surface (Fig. 11a) and the bulk material (Fig. 11b) [71]. In this case, the much faster cooling rate in the surface of the AM Ti-6Al-4V has contributed to the formation of the acicular, metastable martensite phases in Fig. 11(a) while the bulk material presents a bimodal, equilibrium microstructure due to a much lower cooling rate, see Fig.11(b).

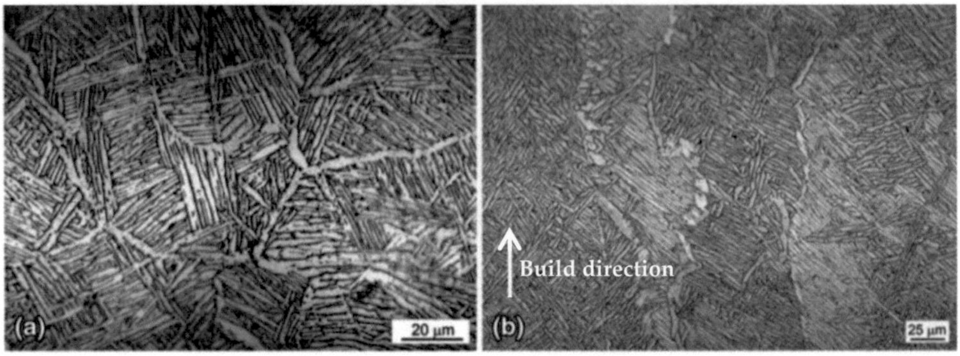

Figure 10. Optical microscopy images of the EBM Ti-6Al-4V from the (a) transverse cross-section and (b) longitudinal cross-section. Build direction is given in (b). Texture/directional growth can be found in (b) [4].

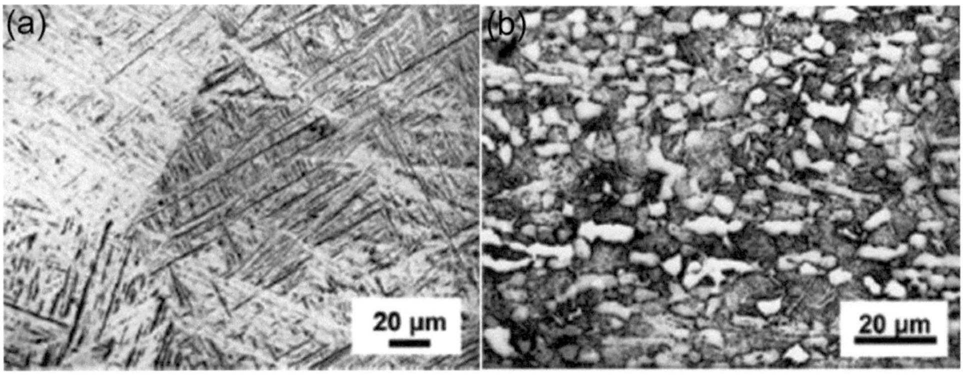

Figure 11. SEM images of the LMD Ti-6Al-4V which show heterogeneity among (a) the surface of the alloy (lamellar with martensite phases) and (b) base/bulk material [71].

5.2.3. Martensite phase transformation

Cooling rate of SLM, EBM and LMD can be 10^4 K/s-10^6 K/s high [29-76]. This enables martensite phases to form in the microstructure. Figure 12 proposes the dependency of phase selection on the cooling rate for Ti-6Al-4V [92,93]. Acicular α, massive martensite α_m or equilibrium (α +β) is proposed for each representative cooling rate. A minimum cooling rate of 20°C/s is suggested to be necessary for the formation of the martensite phases while when cooling rate is higher than 525°C/s the entire microstructure can be featured as acicular α martensite. One still needs to note that the real cooling rate during AM processing of Ti-6Al-4V, other materials as well, is still a research question to be further investigated. Figure 13 provides an example from an EBM Ti-6Al-4V where featherless α can be found in the microstructure [94]. The stable

α phase is also observable in the microstructure although it is suggested that the cooling rate during EBM can be much higher than the critical cooling rate for the martensite phase transformation and accordingly the microstructure should have been overwhelmingly martensite. This is not consistent with the microstructural observation.

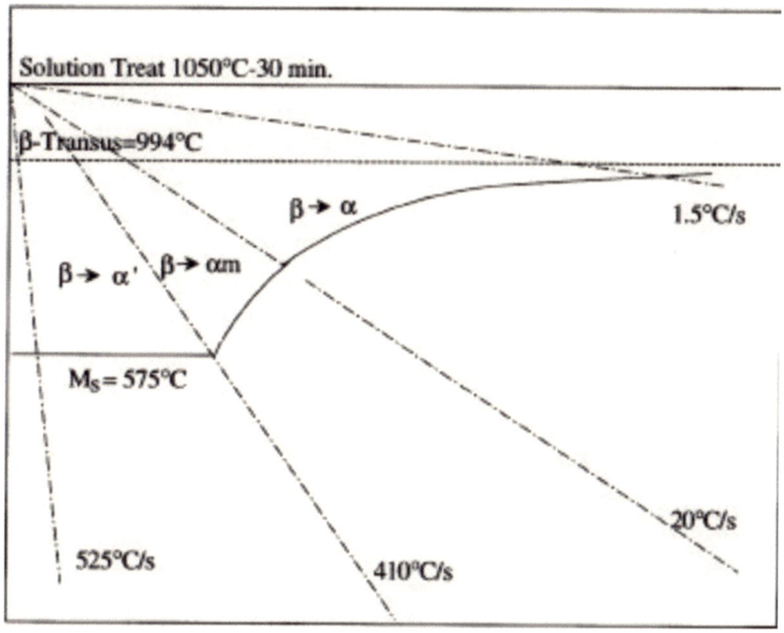

Figure 12. Schematic graph to show the relationship between phase selection and cooling rate during solidification (from 1050°C). The initial state of the Ti-6Al-4V alloy is as β phase [92,93].

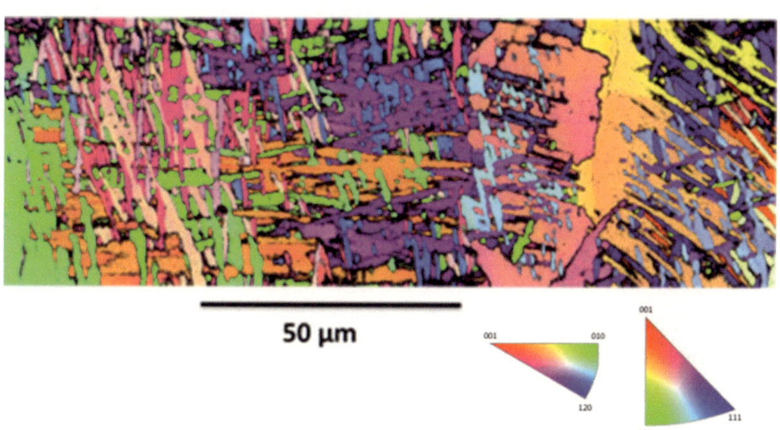

Figure 13. Electron beam scattered diffraction (EBSD) image to show the featureless α' martensite phase and the surrounding microstructure of an EBM Ti-6Al-4V [94].

6. Mechanical property of AM Ti-6Al-4V (As built)

UTS and elongation of as-built AM Ti-6Al-4V are summarised into Figure 14 (a) and (b) [29-88]. ASTM requires that as-prepared Ti-6Al-4V should be no lower than 860 MPa in UTS and no lower than 10% in elongation. These two benchmark values are marked in the two figures using dotted lines. These data suggest that, comparatively speaking, EBM is able to provide a combination of good facture strength and good ductility which satisfy the ASTM specifications. Ti-6Al-4V made by SLM tends to show the highest fracture strength among the four AM approaches yet the corresponding ductility is the lowest which is mostly below the corresponding benchmark value (10%). There are no adequate data for LMD but it is reckoned that it should show a similar tendency to that of the SLM due to the similarity between the two processing techniques. Because of poor density (see Figure 6), SLS Ti-6Al-4V normally relies on post treatment such as HIP to achieve good fracture strength as well as ductility.

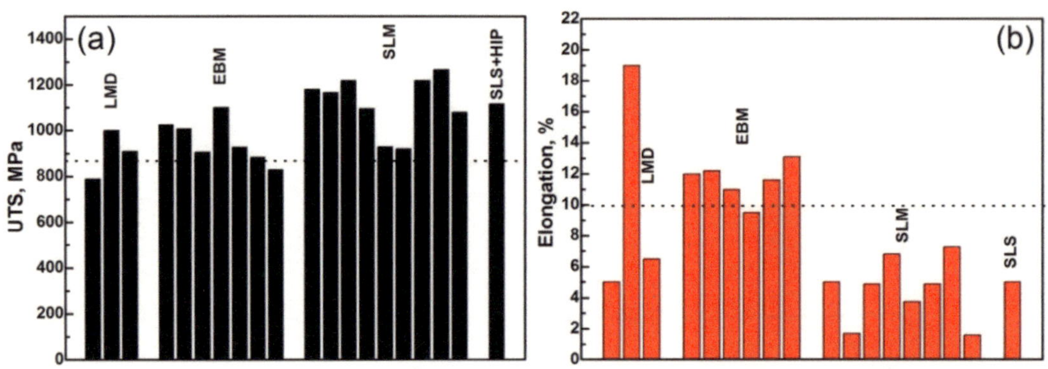

Figure 14. (a) UTS and (b) elongation of as-built Ti-6Al-4V prepared by the various AM techniques [29-88]. Dotted lines in the two figures represent corresponding ASTM specifications.

7. Heat treatment of AM Ti-6Al-4V and corresponding microstructure and mechanical properties

7.1. Heat treatment temperature: A collection from AM Ti-6Al-4V and corresponding mechanical property

Results show that the residual stress in AM Ti-6Al-4V can be higher than 1000 MPa [33]. Considering that the yield strength of Ti-6Al-4V is merely around 880 MPa (see Table 2), residual stress can initiate premature failure especially during dynamic testing such as fatigue. It is a major concern for the AM Ti-6Al-4V. Most as-built AM Ti-6Al-4V therefore need heat treatment to mitigate the impact of the residual stress on the mechanical property performance. A collection of heat treatment temperatures being adopted by current studies is shown in Fig. 15 [29-88]. Temperatures of the martensite phase transformation (at ~610°C), $\alpha \rightarrow \beta$ (at ~750°C) and β-transus (at ~980°C) are also marked in the figure using dotted lines. The figure indicates that most of the heat treatments are distributed at temperatures between 610°C and 980°C. This is very likely due to the necessity of eliminating the hard-but-brittle martensite phases via phase transformation from the α' and/or the α_m phases to stable (α and/or β) phases for the LMD, EBM and SLM approaches. The microstructure of the SLS Ti-6Al-4V is similar to that of the equilibrium state, and can be free from any post heat treatment.

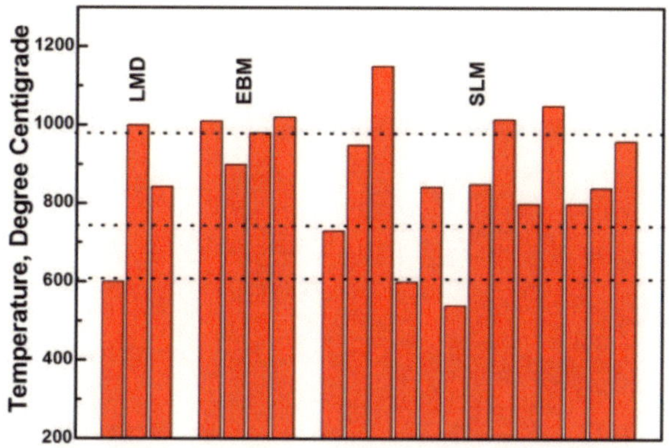

Figure 15. Collection of heat treatment temperatures adopted by various studies of the AM Ti-6Al-4V [29-88]. Dotted lines in the figure represent critical temperatures for Ti-6Al-4V (refer to Table 1).

UTS and elongation of the as-annealed AM Ti-6Al-4V are shown in Fig.16 [29-88]. If comparing with the as-built data (see Fig. 14), the general trend is that the fracture strength will be reduced after heat treatment while ductility can be improved. This is mainly resulted from both the reduced residual stress due to the heat treatment and the partially or even fully transformed martensite phases in the as-built microstructure.

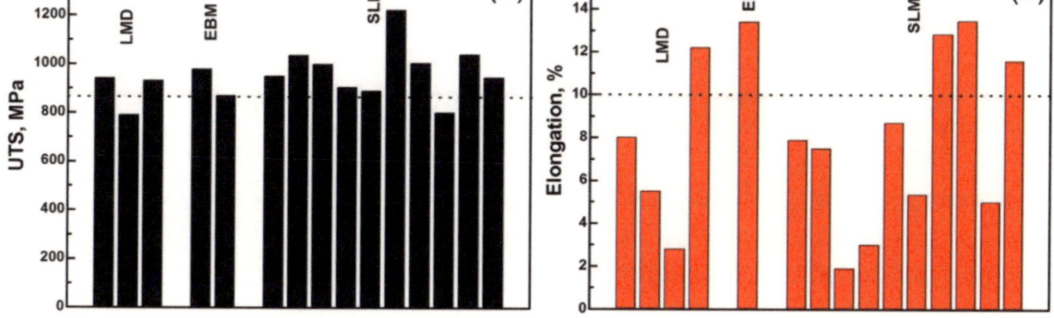

Figure 16. (a) UTS and (b) elongation of as-annealed AM Ti-6Al-4V [29-88]. Dotted lines in the two figures represent corresponding ASTM specifications.

7.2. Stress relief annealing and corresponding microstructure of as-annealed AM Ti-6Al-4V

Heat treatment at temperatures below the temperature of the martensite phase transformation may be not able to change the microstructure. Figure 17 provides such an example via an EBM Ti-6Al-4V annealed at 600°C for four hours [69]. The martensite phases are still observable after the annealing. Such heat treatment may only offer stress relief to reduce the residual stress in the as-built AM Ti-6Al-4V.

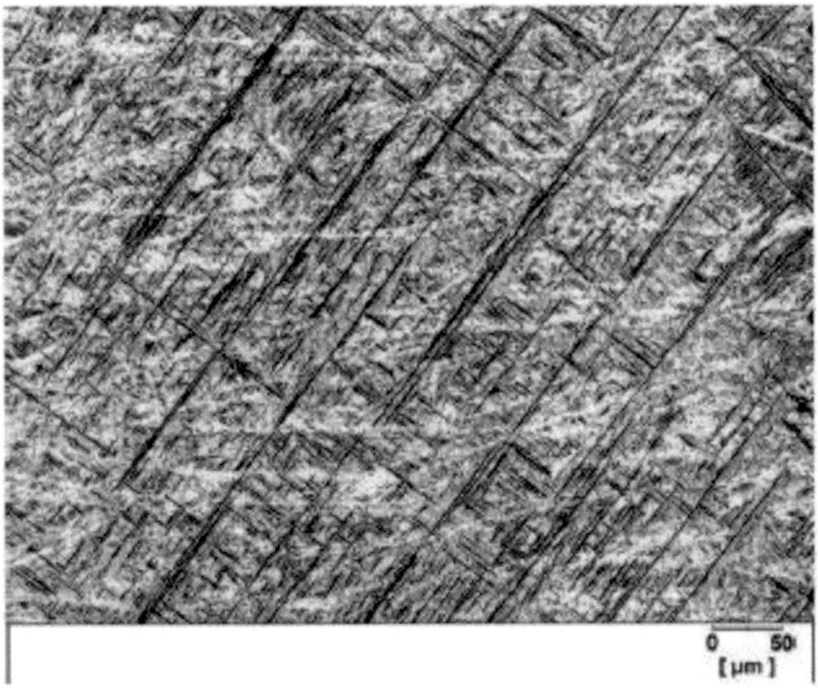

Figure 17. Microstructure of an EBM Ti-6Al-4V annealed at 600°C for four hours. Rectangular (acicular) shaped martensite phases remained after the heat treatment [69].

7.3. Heat treatment to achieve equilibrium microstructure of AM Ti-6Al-4V

7.3.1. Heat treatment at temperature above the martensite phase transformation

Contrary to the low temperature annealing, heat treatment at temperatures higher than the martensite phase transformation may assist in eliminating the brittle, metastable martensite phases via phase transformation [29-88]. Increasing the heat treatment temperature to higher than the $\alpha \rightarrow \beta$ (at ~750°C) or even the β-transus (at ~980°C) will facilitate the transformation from α to β. Figure 18 provides an example of this where heat treatment was conducted up to 1200°C to form equilibrium lamellar (α+β) microstructure [69]. Martensite phases were not observable after such high-temperature heat treatment.

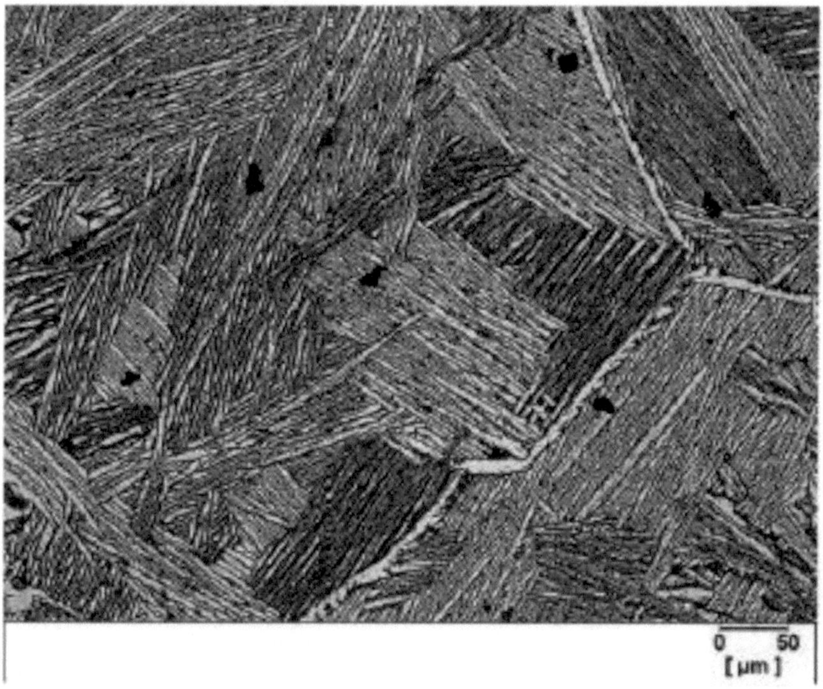

Figure 18. Microstructure of an LMD Ti-6Al-4V annealed at 1200°C for two hours. Colony α-phase is observable along with prior β grains and grain boundary α phases [69].

7.3.2. Dependence of α colony size on heat treatment and microstructural dependency of mechanical property

The model developed by Tiley et al. [95,96] suggests that the mechanical performance of the (α+β) dual phase Ti-6Al-4V can be estimated based on a variety of microstructural parameters including the size of the α colony. The schematic graph in Figure 19 suggests that increasing size of the α colony may lead to reduced ductility (ε) and yield strength ($\sigma_{0.2}$) but in the meantime it may also contribute to better resistance to macro-cracks [95,96]. It is noted that heat treatment can coarsen the α colony of the AM Ti-6Al-4V, see Fig. 20 [54]. From this perspective, heat treatment of AM Ti-6Al-4V will have to be selected according to specific requirement and serve the application purpose of the material.

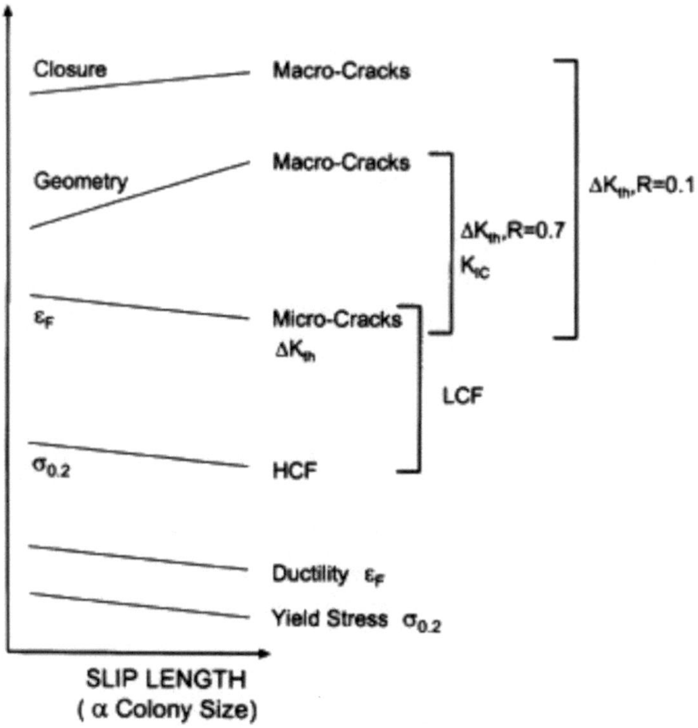

Figure 19. Dependency of mechanical property on the α colony size of the Ti-6Al-4V alloy [95,96].

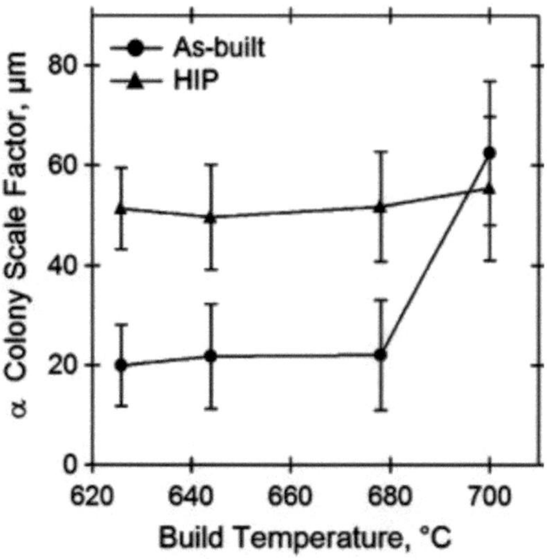

Figure 20. The size of the α colony in an SLS Ti-6Al-4V. HIP at (626°C-700°C) increases the α colony size evidently. HIP at higher temperatures can slightly increase the α colony size than those of lower temperatures [54].

8. Potential oxygen issue for the AM Ti-6Al-4V and counter measurements

8.1. Oxygen equivalent and oxygen level in AM Ti-6Al-4V

Oxygen is detrimental to ductility of Ti-6Al-4V especially when it exceeds certain level [97-99]. Miura et al. [26] suggests the critical oxygen level for PM Ti-6Al-4V is 0.33 wt.%, above which ductility drops rapidly and can be much lower than the corresponding ASTM specification (see Fig. 1b). It needs to be noted that C, N and Fe may affect the mechanical performance of Ti-6Al-4V in a similar way to that of the oxygen, as suggested by the following Eq. (3) [100] and Eq.(4) [101].

$$[O]_{EQ} = [O] + 2[N] + 2/3[C] \ (\text{wt.\%}) \tag{3}$$

$$[O]_{EQ} = [O] + 2.77[N] + 0.1[Fe] \ (\text{wt.\%}) \tag{4}$$

Generally speaking, AM processing only mildly increases oxygen level and the oxygen level in most of the as-built AM Ti-6Al-4V remain low [29-88]. This is mostly contributed to the extra-low interstitial (ELI) raw powders used, good vacuum condition and/or inert gas protection achievable during AM processing. Only occasional report can be found where oxygen level is higher than the 0.2 wt.%O benchmark value, see Figure 21 [29-88].

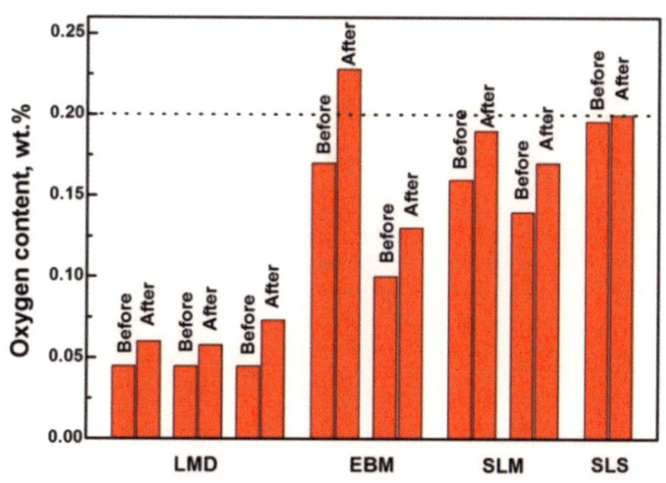

Figure 21. Variation of oxygen content before and after AM [29-88]. Dotted line in the figure represents corresponding ASTM specification (0.2 wt.%O).

8.2. Price issue of raw AM powders

Figure 21 indicates that the oxygen issue may have not been serious to the AM Ti-6Al-4V so far. This is mainly thanks to the use of ELI powders that are prepared by gas-atomisation or rotating electrode. The high cost of the ELI raw powders, however, can be an issue to AM Ti-6Al-4V from cost perspective. Figure 22 compares the price of a variety of powders, showing that the cost on the AM powders is extraordinary [4-13]. They are mostly higher than $400 per kg while hydrogenation-dehydrogenation (HDH) Ti powders that are widely used in conventional PM Ti are merely around $30 per kg. The high cost can be the bottle-neck issue for AM Ti-6Al-4V to be fully embraced by industries. For this reason, low-cost, high-interstitial (i.e. oxygen), fine powders are more desired for developing cost-effective AM Ti alloys. Sun et al. [102] recently developed an approach to manipulate irregularly-shaped HDH powders to make round-shaped, fine particles that are suitable for AM. The cost of their powders is higher than the HDH Ti powder but still markedly lower than the current AM powders, making it be more reliable for developing AM Ti and Ti alloys.

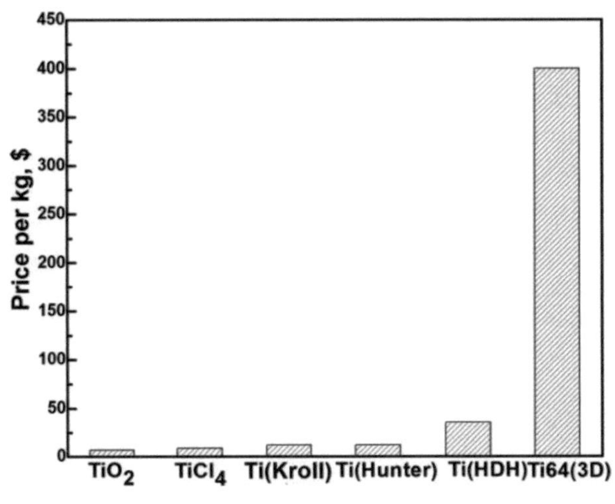

Figure 22. Comparison on the cost of various Ti materials, including raw material (TiO_2), intermediate material ($TiCl_4$) and Ti powders produced via the Kroll, Hunter and HDH approaches (price of Ti-6Al-4V is similar to these Ti powders). Price of fine powders of Ti-6Al-4V for the 3D printing is extraordinary [4-13].

8.3. Counter measurements

Counter measurements need to be consulted when the high impurity (oxygen) powders are to be used for developing AM Ti-6Al-4V. Research has shown that the RE elements are capable to scavenge oxygen and their oxygen scavenging capability follows this sequence: Y>Er>Dy>Tb>Gd (see Figure 23a) [103, 104]. The high potency of yttrium in scavenging oxygen from AM titanium alloys is demonstrated in Figure 23 (b), where the uniformly distributed Y_2O_3 dispersoids are resulted from an addition of 0.1 wt.%Y to an EBM titanium alloy [105].

RE hydrides such as YH_2 have been further demonstrated to be able to scavenge, aside from oxygen, Cl, another important impurity element to Ti materials, Fig. 24 [106,107].

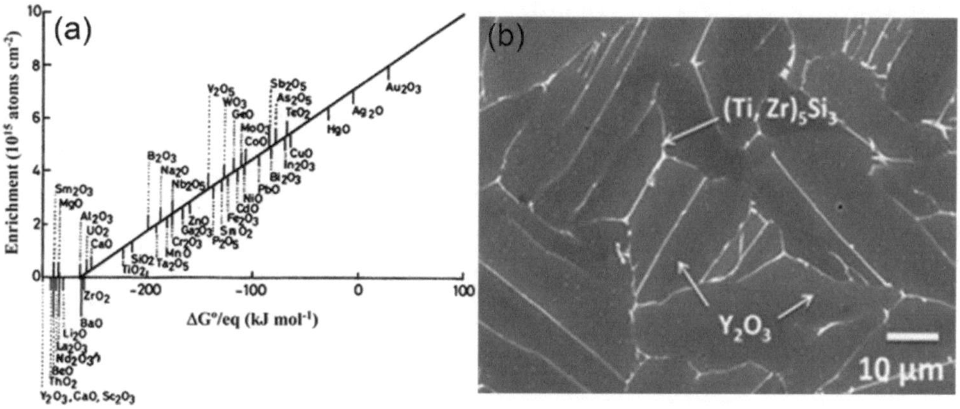

Figure 23. The free Gipps energy (ΔG) of formation of various oxide materials, showing their thermodynamic stability as well as their affinity for oxygen [104]. Rare earth elements such as Y and some alkaline earth elements such as Ca have higher affinity for oxygen than Ti. (b) Precipitation of fine Y_2O_3 dispersoids (due to the addition of 0.1 wt.%Y) in the α-Ti matrix of Ti-6Al-2.7Sn-4Zr-0.4Mo-0.45Si-0.1Y (wt.%) which contained 0.07 wt.% oxygen. The alloy was additively manufactured by EBM [105].

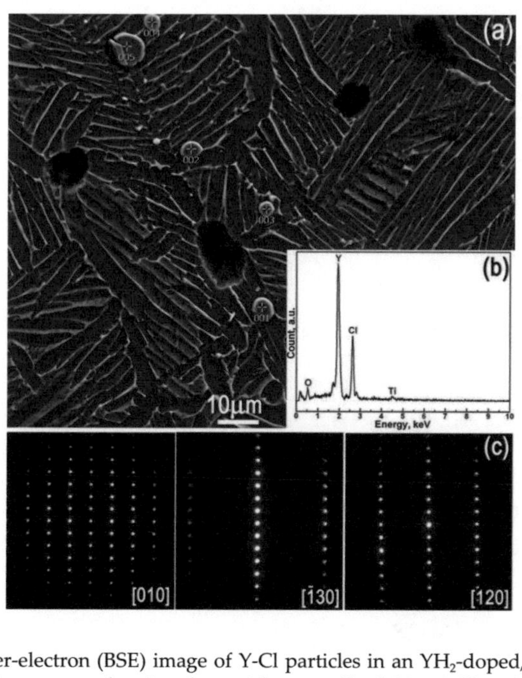

Figure 24. (a) SEM back-scatter-electron (BSE) image of Y-Cl particles in an YH_2-doped, as-sintered Ti alloy, (b) SEM energy dispersive x-ray (EDX) spectrum showing the enrichment of both Y and Cl in the phase and (c) TEM selected area electron diffraction (SAED) analysis of the phase [107].

9. Concluding remarks

3D printing has becoming a focusing topic not only to research community and industry but also to the general public, and AM Ti and Ti alloys is one of the most promising and interested areas to be further developed. For the time being, although a few issues persist such as the microstructural inhomogeneity in the as-built material, some of the AM Ti-6Al-4V have already been able to achieve mechanical properties no lower than the corresponding ASTM specifications. The cost of the ELI AM Ti powders is one of the most challenging issues to limit the scale-up of the AM products. Employing low-cost powders to replace existing expensive powders to reduce the overall cost is a valuable research direction for further developing AM Ti-6Al-4V. Counter measurements to deal with the impurity issue associated with the high-interstitial Ti powders can be one of the key research elements for such development.

Author details

Ming Yan[1,2*] and Peng Yu[2]

*Address all correspondence to: yan.m@sustc.edu.cn

1 RMIT University, School of Aerospace, Mechanical and Manufacturing and Centre of Additive Manufacture, Melbourne, VIC, Australia

2 The South University of Science and Technology, China

References

[1] Kobryn P A, Semiatin S L. The laser additive manufacture of Ti-6Al-4V. JOM, 2001, 53(9): 40-42.

[2] Chahine, G., Koike, M., Okabe, T., Smith, P., & Kovacevic, R. (2008). The design and production of Ti-6Al-4V ELI customized dental implants. JOM, 60(11), 50-55.

[3] Ivanova O, Williams C, Campbell T. Additive manufacturing (AM) and nanotechnology: promises and challenges. Rapid Prototyping Journal, 2013, 19(5): 353-364.

[4] Froes F H, Eylon D. Powder metallurgy of titanium alloys. International Materials Reviews, 1990, 35(1): 162-184.

[5] Imam M A, Froes F H S. Low cost titanium and developing applications. JOM, 2010, 62(5): 17-20.

[6] Qian, M. (2010). Cold compaction and sintering of titanium and its alloys for near-net-shape or preform fabrication. International journal of powder metallurgy, 46(5), 29-44.

[7] Ivasishin, O. M., Anokhin, V. M., Demidik, A. N., & Savvakin, D. G. (2000). Cost-effective blended elemental powder metallurgy of titanium alloys for transportation application. Key Engineering Materials, 188, 55-62.

[8] Qian, M., Yang, Y.F., Luo, S.D., Tang, H.P., Pressureless sintering of titanium and titanium alloys: sintering densification and solute hominization. In "Titanium Powder Metallurgy", (Editors) Qian, M., Froes, F.H. Elsevier and Science Direct, ISBN: 978-0-12-800054-0, in press, DOI: 10.1007/s11661-014-2631-4, 2014.

[9] Wang, Hongtao, Z. Zak Fang, and Pei Sun. A critical review of mechanical properties of powder metallurgy titanium. International journal of powder metallurgy 46, no. 5 (2010): 45-57.

[10] Yan, M., S. D. Luo, G. B. Schaffer, and M. Qian. TEM and XRD characterisation of commercially pure α-Ti made by powder metallurgy and casting. Materials Letters 72 (2012): 64-67.

[11] Yan, M., S. D. Luo, G. B. Schaffer, and M. Qian. Impurity (Fe, Cl, and P)-Induced Grain Boundary and Secondary Phases in Commercially Pure Titanium (CP-Ti). Metallurgical and Materials Transactions A 44, no. 8 (2013): 3961-3969.

[12] Yan, M., Qian, M., Kong, C., & Dargusch, M. S. (2014). Impacts of trace carbon on the microstructure of as-sintered biomedical Ti–15Mo alloy and reassessment of the maximum carbon limit. Acta biomaterialia, 10(2), 1014-1023.

[13] Yan, M., Dargusch, M.S., Kong, C., Kimpton, J., Kohara, S., Brandt, M., & Qian, M., In-situ synchrotron x-ray diffraction of the powder compacts of TiH_2-6Al-4V and Ti-6Al-4V: Accelerated alloying and phase transformation and the formation of an oxygen-enriched Ti_4Fe_2O phase in TiH_2-6Al-4V, Metallurgical and Materials Transactions A, in press, 2014.

[14] Leyens, C., & Peters, M. (2003). Titanium and titanium alloys (p. 187). Wiley-VCH, Weinheim.

[15] Lütjering, G., & Williams, J. C. (2003). Titanium (Vol. 2). Berlin: Springer

[16] Yang, Y. F., Imai, H., Kondoh, K., & Qian, M. (2014). Comparison of spark plasma sintering of elemental and master alloy powder mixes and prealloyed Ti-6Al-4V powder. International Journal of Powder Metallurgy, 50(1), 41-47.

[17] Lenel, F. V. (1980). Powder metallurgy: principles and applications (Vol. 225). Princeton, NJ: Metal Powder Industries Federation.

[18] German, R. M. (2005). Powder metallurgy & particulate materials processing (pp. 386-387). Princeton, NJ: Metal powder industries federation.

[19] Kingery, W. D., Bowen, H. K., & Uhlmann, D. R. Introduction to ceramics, 1976. New York, 788-1016.

[20] Crosby, K. D. (2013). Titanium-6Aluminum-4Vanadium for functionally graded orthopedic implant applications, Doctoral thesis, University of Connecticut.

[21] Robertson, I. M., & Schaffer, G. B. (2010). Comparison of sintering of titanium and titanium hydride powders. Powder Metallurgy, 53(1), 12-19.

[22] Fang, Z. Z., Sun, P., & Wang, H. (2012). Hydrogen Sintering of Titanium to Produce High Density Fine Grain Titanium Alloys. Advanced Engineering Materials, 14(6), 383-387.

[23] Li, S., Sun, B., Imai, H., & Kondoh, K. (2013). Powder metallurgy Ti–TiC metal matrix composites prepared by in situ reactive processing of Ti-VGCFs system. Carbon, 61, 216-228.

[24] Yan, M., Microstructural characterization of as-sintered Ti and Ti alloys. In "Titanium Powder Metallurgy", (Editors) Qian, M., Froes, F.H. Elsevier and Science Direct, ISBN: 978-0-12-800054-0, In press, 2014.

[25] Yan, M., M. S. Dargusch, T. Ebel, and M. Qian. A transmission electron microscopy and three-dimensional atom probe study of the oxygen-induced fine microstructural features in as-sintered Ti–6Al–4V and their impacts on ductility. Acta Materialia 68 (2014): 196-206.

[26] Miura, H., Itoh, Y., Ueamtsu, T., & Sato, K. (2010). The influence of density and oxygen content on the mechanical properties of injection molded Ti-6Al-4V alloys. Advances in Powder Metallurgy and Particulate Materials, 46-53.

[27] http://asm.matweb.com/search/SpecificMaterial.asp?bassnum=MTP641

[28] Elmer, J. W., Palmer, T. A., Babu, S. S., Zhang, W., & DebRoy, T. (2004). Phase transformation dynamics during welding of Ti–6Al–4V. Journal of applied physics, 95(12), 8327-8339.

[29] Roberts, I. A., Wang, C. J., Esterlein, R., Stanford, M., & Mynors, D. J. (2009). A three-dimensional finite element analysis of the temperature field during laser melting of metal powders in additive layer manufacturing. International Journal of Machine Tools and Manufacture, 49(12), 916-923.

[30] Facchini, L., Magalini, E., Robotti, P., Molinari, A., Höges, S., & Wissenbach, K. (2010). Ductility of a Ti-6Al-4V alloy produced by selective laser melting of prealloyed powders. Rapid Prototyping Journal, 16(6), 450-459.

[31] Thijs, L., Verhaeghe, F., Craeghs, T., Humbeeck, J. V., & Kruth, J. P. (2010). A study of the microstructural evolution during selective laser melting of Ti–6Al–4V. Acta Materialia, 58(9), 3303-3312.

[32] Vilaro, T., Colin, C., & Bartout, J. D. (2011). As-fabricated and heat-treated microstructures of the Ti-6Al-4V alloy processed by selective laser melting. Metallurgical and Materials Transactions A, 42(10), 3190-3199.

[33] Knowles, C.R., Becker, T.H., Tait, R.B., Residual stress measurement and structural integrity implications for selective laser melted Ti-6Al-4V. South African Journal of Industrial Engineering, 2012, 23(3), 119-129.

[34] Pyka, G., Burakowski, A., Kerckhofs, G., Moesen, M., Van Bael, S., Schrooten, J., & Wevers, M. (2012). Surface modification of Ti6Al4V open porous structures produced by additive manufacturing. Advanced Engineering Materials, 14(6), 363-370.

[35] SU, X. B., YANG, Y. Q., YU, P., & SUN, J. F. (2012). Development of porous medical implant scaffolds via laser additive manufacturing. Transactions of Nonferrous Metals Society of China, 22, s181-s187.

[36] Lin, C. W., Ju, C. P., & Chern Lin, J. H. (2005). A comparison of the fatigue behavior of cast Ti–7.5 Mo with cp titanium, Ti–6Al–4V and Ti–13Nb–13Zr alloys. Biomaterials, 26(16), 2899-2907.

[37] Vrancken, B., Thijs, L., Kruth, J. P., & Van Humbeeck, J. (2012). Heat treatment of Ti6Al4V produced by Selective Laser Melting: Microstructure and mechanical properties. Journal of Alloys and Compounds, 541, 177-185.

[38] Leuders, S., Thöne, M., Riemer, A., Niendorf, T., Tröster, T., Richard, H. A., & Maier, H. J. (2013). On the mechanical behaviour of titanium alloy TiAl6V4 manufactured by selective laser melting: Fatigue resistance and crack growth performance. International Journal of Fatigue, 48, 300-307.

[39] Van Bael, S., Chai, Y. C., Truscello, S., Moesen, M., Kerckhofs, G., Van Oosterwyck, H.,... & Schrooten, J. (2012). The effect of pore geometry on the in vitro biological behavior of human periosteum-derived cells seeded on selective laser-melted Ti6Al4V bone scaffolds. Acta biomaterialia, 8(7), 2824-2834.

[40] Qiu, C., Adkins, N. J., & Attallah, M. M. (2013). Microstructure and tensile properties of selectively laser-melted and of HIPed laser-melted Ti–6Al–4V. Materials Science and Engineering: A, 578, 230-239.

[41] Pyka, G., Kerckhofs, G., Papantoniou, I., Speirs, M., Schrooten, J., & Wevers, M. (2013). Surface roughness and morphology customization of additive manufactured open porous Ti6Al4V structures. Materials, 6(10), 4737-4757.

[42] Chan, K. S., Koike, M., Mason, R. L., & Okabe, T. (2013). Fatigue Life of Titanium Alloys Fabricated by Additive Layer Manufacturing Techniques for Dental Implants. Metallurgical and Materials Transactions A, 44(2), 1010-1022.

[43] Amin Yavari, S., Wauthlé, R., van der Stok, J., Riemslag, A. C., Janssen, M., Mulier, M.,... & Zadpoor, A. A. (2013). Fatigue behavior of porous biomaterials manufactured using selective laser melting. Materials Science and Engineering: C, 33(8), 4849-4858.

[44] Rafi, H. K., Karthik, N. V., Gong, H., Starr, T. L., & Stucker, B. E. (2013). Microstructures and Mechanical Properties of Ti6Al4V Parts Fabricated by Selective Laser Melting and Electron Beam Melting. Journal of materials engineering and performance, 22(12), 3872-3883.

[45] Yadroitsev, I., Krakhmalev, P., & Yadroitsava, I. (2014). Selective laser melting of Ti6Al4V alloy for biomedical applications: Temperature monitoring and microstructural evolution. Journal of Alloys and Compounds, 583, 404-409.

[46] http://compolight.dti.dk/29509,2 (Courtesy of Olivier Jay)

[47] Nakamura, H., & Mitchell, A. (1992). The effect of beam oscillation rate on Al evaporation from a Ti-6Al-4V alloy in the electron beam melting process. ISIJ international, 32(5), 583-592.

[48] Zhuk, H. V., Kobryn, P. A., & Semiatin, S. L. (2007). Influence of heating and solidification conditions on the structure and surface quality of electron-beam melted Ti–6Al–4V ingots. Journal of materials processing technology, 190(1), 387-392.

[49] Kalinyuk, A. N., Trigub, N. P., Zamkov, V. N., Ivasishin, O. M., Markovsky, P. E., Teliovich, R. V., & Semiatin, S. L. (2003). Microstructure, texture, and mechanical properties of electron-beam melted Ti–6Al–4V. Materials Science and Engineering: A, 346(1), 178-188.

[50] Heinl, P., Müller, L., Körner, C., Singer, R. F., & Müller, F. A. (2008). Cellular Ti–6Al–4V structures with interconnected macro porosity for bone implants fabricated by selective electron beam melting. Acta biomaterialia, 4(5), 1536-1544.

[51] Facchini, L., Magalini, E., Robotti, P., & Molinari, A. (2009). Microstructure and mechanical properties of Ti-6Al-4V produced by electron beam melting of pre-alloyed powders. Rapid Prototyping Journal, 15(3), 171-178.

[52] Ponader, S., Von Wilmowsky, C., Widenmayer, M., Lutz, R., Heinl, P., Körner, C.,... & Schlegel, K. A. (2010). In vivo performance of selective electron beam-melted Ti-6Al-4V structures. Journal of biomedical materials research Part A, 92(1), 56-62.

[53] Murr, L. E., Gaytan, S. M., Medina, F., Martinez, E., Martinez, J. L., Hernandez, D. H.,... & Wicker, R. B. (2010). Characterization of Ti–6Al–4V open cellular foams fabricated by additive manufacturing using electron beam melting. Materials Science and Engineering: A, 527(7), 1861-1868.

[54] Al-Bermani, S. S., Blackmore, M. L., Zhang, W., & Todd, I. (2010). The Origin of Microstructural Diversity, Texture, and Mechanical Properties in Electron Beam Melted Ti-6Al-4V. Metallurgical and materials transactions a, 41(13), 3422-3434.

[55] Murr, L. E., Amato, K. N., Li, S. J., Tian, Y. X., Cheng, X. Y., Gaytan, S. M.,... & Wicker, R. B. (2011). Microstructure and mechanical properties of open-cellular biomaterials prototypes for total knee replacement implants fabricated by electron beam melting. Journal of the mechanical behavior of biomedical materials, 4(7), 1396-1411.

[56] Cronskär, M., Bäckström, M., & Rännar, L. E. (2013). Production of customized hip stem prostheses–a comparison between conventional machining and electron beam melting (EBM). Rapid Prototyping Journal, 19(5), 365-372.

[57] Li, S. J., Murr, L. E., Cheng, X. Y., Zhang, Z. B., Hao, Y. L., Yang, R.,... & Wicker, R. B. (2012). Compression fatigue behavior of Ti–6Al–4V mesh arrays fabricated by electron beam melting. Acta Materialia, 60(3), 793-802.

[58] Bush, R. W., & Brice, C. A. (2012). Elevated temperature characterization of electron beam freeform fabricated Ti–6Al–4V and dispersion strengthened Ti–8Al–1Er. Materials Science and Engineering: A, 554, 12-21.

[59] Cheng, X. Y., Li, S. J., Murr, L. E., Zhang, Z. B., Hao, Y. L., Yang, R.,... & Wicker, R. B. (2012). Compression deformation behavior of Ti–6Al–4V alloy with cellular structures fabricated by electron beam melting. Journal of the mechanical behavior of biomedical materials, 16, 153-162.

[60] Yang, L., Harrysson, O., West, H., & Cormier, D. (2012). Compressive properties of Ti–6Al–4V auxetic mesh structures made by electron beam melting. Acta Materialia, 60(8), 3370-3379.

[61] Safdar, A., Wei, L. Y., Snis, A., & Lai, Z. (2012). Evaluation of microstructural development in electron beam melted Ti-6Al-4V. Materials Characterization, 65, 8-15.

[62] Karlsson, J., Snis, A., Engqvist, H., & Lausmaa, J. (2013). Characterization and comparison of materials produced by Electron Beam Melting (EBM) of two different Ti–6Al–4V powder fractions. Journal of Materials Processing Technology, 213(12), 2109-2118.

[63] Antonysamy, A. A., Meyer, J., & Prangnell, P. B. (2013). Effect of build geometry on the β-grain structure and texture in additive manufacture of Ti6Al4V by selective electron beam melting. Materials Characterization, 84, 153-168.

[64] Hrabe, N., & Quinn, T. (2013). Effects of processing on microstructure and mechanical properties of a titanium alloy (Ti–6Al–4V) fabricated using electron beam melting (EBM), part 1: Distance from build plate and part size. Materials Science and Engineering: A, 573, 264-270.

[65] Hrabe, N., & Quinn, T. (2013). Effects of processing on microstructure and mechanical properties of a titanium alloy (Ti–6Al–4V) fabricated using electron beam melting (EBM), Part 2: Energy input, orientation, and location. Materials Science and Engineering: A, 573, 271-277.

[66] Schwerdtfeger, Jan, Robert F. Singer, and Carolin Körner. In situ flaw detection by IR-imaging during electron beam melting." Rapid Prototyping Journal 18, no. 4 (2012): 259-263.

[67] Banerjee, R., Collins, P. C., Genc, A., & Fraser, H. L. (2003). Direct laser deposition of in situ Ti–6Al–4V–TiB composites. Materials Science and Engineering: A, 358(1), 343-349.

[68] Miranda, R. M., Lopes, G., Quintino, L., Rodrigues, J. P., & Williams, S. (2008). Rapid prototyping with high power fiber lasers. Materials & Design, 29(10), 2072-2075.

[69] Brandl, E., Palm, F., Michailov, V., Viehweger, B., & Leyens, C. (2011). Mechanical properties of additive manufactured titanium (Ti–6Al–4V) blocks deposited by a solid-state laser and wire. Materials & Design, 32(10), 4665-4675.

[70] Brandl, E., Michailov, V., Viehweger, B., & Leyens, C. (2011). Deposition of Ti–6Al–4V using laser and wire, part I: Microstructural properties of single beads. Surface and Coatings Technology, 206(6), 1120-1129.

[71] Brandl, E., Michailov, V., Viehweger, B., & Leyens, C. (2011). Deposition of Ti–6Al–4V using laser and wire, part II: Hardness and dimensions of single beads. Surface and Coatings Technology, 206(6), 1130-1141.

[72] Baufeld, B., Brandl, E., & Van der Biest, O. (2011). Wire based additive layer manufacturing: comparison of microstructure and mechanical properties of Ti–6Al–4V components fabricated by laser-beam deposition and shaped metal deposition. Journal of Materials Processing Technology, 211(6), 1146-1158.

[73] Chen, J., Xue, L., & Wang, S. H. (2011). Experimental studies on process-induced morphological characteristics of macro-and microstructures in laser consolidated alloys. Journal of materials science, 46(17), 5859-5875.

[74] Ahsan, M. N., Pinkerton, A. J., Moat, R. J., & Shackleton, J. (2011). A comparative study of laser direct metal deposition characteristics using gas and plasma-atomized Ti–6Al–4V powders. Materials Science and Engineering: A, 528(25), 7648-7657.

[75] Brandl, E., Schoberth, A., & Leyens, C. (2012). Morphology, microstructure, and hardness of titanium (Ti-6Al-4V) blocks deposited by wire-feed additive layer manufacturing (ALM). Materials Science and Engineering: A, 532, 295-307.

[76] Brandl, E., & Greitemeier, D. (2012). Microstructure of additive layer manufactured Ti–6Al–4V after exceptional post heat treatments. Materials Letters, 81, 84-87.

[77] Clark, D., Whittaker, M. T., & Bache, M. R. (2012). Microstructural characterization of a prototype titanium alloy structure processed via direct laser deposition (DLD). Metallurgical and Materials Transactions B, 43(2), 388-396.

[78] Das, M., Balla, V. K., Basu, D., Manna, I., Sampath Kumar, T. S., & Bandyopadhyay, A. (2012). Laser processing of in situ synthesized TiB–TiN-reinforced Ti6Al4V alloy coatings. Scripta Materialia, 66(8), 578-581.

[79] Das, S., Wohlert, M., Beaman, J. J., & Bourell, D. L. (1998). Producing metal parts with selective laser sintering/hot isostatic pressing. JoM, 50(12), 17-20.

[80] Das, S., Beama, J. J., Wohlert, M., & Bourell, D. L. (1998). Direct laser freeform fabrication of high performance metal components. Rapid Prototyping Journal, 4(3), 112-117.

[81] Das, S., Wohlert, M., Beaman, J. J., & Bourell, D. L. (1999). Processing of titanium net shapes by SLS/HIP. Materials & design, 20(2), 115-121.

[82] Ramosoeu, M. E., Chikwanda, H. K., Bolokang, A. S., Booysen, G., & Ngonda, T. N. (2010). Additive manufacturing: Characterization of TI-6AI-4V alloy intended for biomedical application.

[83] Bertol, L. S., Júnior, W. K., Silva, F. P. D., & Aumund-Kopp, C. (2010). Medical design: direct metal laser sintering of Ti–6Al–4V. Materials & Design, 31(8), 3982-3988.

[84] Salmi, M., Tuomi, J., Paloheimo, K. S., Björkstrand, R., Paloheimo, M., Salo, J.,... & Mäkitie, A. A. (2012). Patient-specific reconstruction with 3D modeling and DMLS additive manufacturing. Rapid Prototyping Journal, 18(3), 209-214.

[85] Cooper, D. E., Stanford, M., Kibble, K. A., & Gibbons, G. J. (2012). Additive manufacturing for product improvement at Red Bull Technology. Materials & Design, 41, 226-230.

[86] Witek, L., Marin, C., Granato, R., Bonfante, E. A., Campos, F., Bisinotto, J.,... & Coelho, P. G. (2012). Characterization and in vivo evaluation of laser sintered dental endosseous implants in dogs. Journal of Biomedical Materials Research Part B: Applied Biomaterials, 100(6), 1566-1573.

[87] Stübinger, S., Mosch, I., Robotti, P., Sidler, M., Klein, K., Ferguson, S. J., & Rechenberg, B. (2013). Histological and biomechanical analysis of porous additive manufactured implants made by direct metal laser sintering: A pilot study in sheep. Journal of Biomedical Materials Research Part B: Applied Biomaterials, 101(7), 1154-1163.

[88] http://en.wikipedia.org/wiki/Selective_laser_sintering

[89] German, R. M. (1996). Sintering theory and practice. Sintering Theory and Practice, by Randall M. German, pp. 568. ISBN 0-471-05786-X. Wiley-VCH, January 1996., 1.

[90] German, R. M., Suri, P., & Park, S. J. (2009). Review: liquid phase sintering. Journal of materials science, 44(1), 1-39.

[91] Malinov, S., W. Sha, and Z. Guo. Application of artificial neural network for prediction of time–temperature–transformation diagrams in titanium alloys. Materials Science and Engineering: A 283, no. 1 (2000): 1-10.

[92] Ahmed, T., and H. J. Rack. Phase transformations during cooling in α β titanium alloys. Materials Science and Engineering: A 243, no. 1 (1998): 206-211.

[93] Majdic, Mojda, and Giinter Ziegler. Effect of the Metastable Beta-Phase on Phase Transformations in the Ti Alloy TiAl6V4. Z. Metallkunde 64, no. 11 (1973): 751-758.

[94] Lu, S.L., Qian, M., Yan, M., Tang, H.P, & StJohn, D. (2014) Unpublished materials.

[95] Kar, S., T. Searles, E. Lee, G. B. Viswanathan, H. L. Fraser, J. Tiley, and R. Banerjee. Modeling the tensile properties in β-processed α/β Ti alloys. Metallurgical and Materials Transactions A 37, no. 3 (2006): 559-566.

[96] Collins, P. C., B. Welk, T. Searles, J. Tiley, J. C. Russ, and H. L. Fraser. Development of methods for the quantification of microstructural features in α+β-processed α/β titanium alloys. Materials Science and Engineering: A 508, no. 1 (2009): 174-182.

[97] Yan, M., Xu, W., Dargusch, M.S., Tang, H.P., Brand & M., Qian, M. (2014), Powder Metallurgy, 57, no.4(2014): 251-257.

[98] Yan, M., M. S. Dargusch, T. Ebel, and M. Qian. A transmission electron microscopy and three-dimensional atom probe study of the oxygen-induced fine microstructural features in as-sintered Ti–6Al–4V and their impacts on ductility. Acta Materialia 68 (2014): 196-206.

[99] Yan, M., Y. Liu, G. B. Schaffer, and M. Qian. In situ synchrotron radiation to understand the pathways for the scavenging of oxygen in commercially pure Ti and Ti–6Al–4V by yttrium hydride. Scripta Materialia 68, no. 1 (2013): 63-66.

[100] H.R. Ogden, R.L. Jaffee: The effects of carbon, oxygen, and nitrogen on the mechanical properties of titanium and titanium alloys, TML Report No 20 Battelle Memorial Institute, 1995.

[101] Soeda; Seiichi (Tokyo, JP), Fujii; Hideki (Futtsu, JP), Okano; Hiroyuki (Chigasaki, JP), Hanaki; Michio (Chigasaki, JP): US patent, No. 6063211, "High strength, high ductility titanium-alloy and process for producing the same", 2000.

[102] Sun, Y.Y., Gulizia, S., Oh, C.H., Doblin, C., Yang, Y.F., Qian, M. Unpublished data.

[103] Okabe, T. H., Hirota, K., Kasai, E., Saito, F., Waseda, Y., & Jacob, K. T. (1998). Thermodynamic properties of oxygen in RE–O (RE=Gd, Tb, Dy, Er) solid solutions. Journal of alloys and compounds, 279(2), 184-191.

[104] Thompson, G. E., Skeldon, P., Zhou, X., Shimizu, K., Habazaki, H., & Smith, C. J. E. (2003). Improving the performance of aerospace alloys. Aircraft Engineering and Aerospace Technology, 75(4), 372-379.

[105] Tang, H.P., & Qian, M. Unpublished data.

[106] Yan, M., Tang, H.P., Qian, M.. Scavangin of oxygen ad chlorine from powder metallurgy (PM) titanium and titanium alloys. In "Titanium Powder Metallurgy", (Editors) Qian, M., Froes, F.H. Elsevier and Science Direct, ISBN: 978-0-12-800054-0, In press, 2014.

[107] Yan, M., Y. Liu, Y. B. Liu, C. Kong, G. B. Schaffer, and M. Qian. Simultaneous gettering of oxygen and chlorine and homogenization of the β phase by rare earth hydride additions to a powder metallurgy Ti–2.25 Mo–1.5 Fe alloy. Scripta Materialia 67, no. 5 (2012): 491-494.

Chapter 6

Metal Matrix Composites Added of Nanostructured Tantalum Carbide

Leiliane Alves de Oliveira, Uílame Umbelino Gomes,
Ariadne Souza Silva and Emanuel João Batista

Additional information is available at the end of the chapter

http://dx.doi.org/10.5772/59187

1. Introduction

Metal matrix composites (MMC) reinforced with dispersed ceramic particles have received considerable interest over the years and are still in constant development in order to expand their applications in industry. It is a brilliant choice for applications that require mechanical strength and wear resistance. They combine a soft metal matrix with hard ceramic particles resistant to wear (Gordo, 200).

Different matrixes and reinforcements for MMC have been studied, and therefore studies have been conducted with various combinations of metal matrixes with reinforcement of ceramic powders aimed at obtaining composites with similar properties or superior to conventional steel tool. Carbides such as NbC, TaC, VC, and TiC have been combined with iron or steel powders to produce sintered composites. Niobium and tantalum carbide can be used for structural purposes and also for the production of refractory components, or as steel reinforcement by dispersed particles. Particularly, reinforced steel has been used in the automotive and textile industries and also in the manufacture of high-speed cutting and ore milling tools (Martinelli, *et al*, 2007). Advances in research (Silva, 2005; Silva, *et al*, 2005; and Silva, *et al*, 2012) and marketing of steel reinforced with NbC and TaC particles have also contributed to add value to manufactured products strategically using mineral sources produced in northeastern Brazil.

New techniques for the production of refractory metal carbides (WC, NbC, TaC, TaxNby) have been developed by synthesizing nanostructured carbides that provide improvement of diverse properties of materials compared to materials obtained by conventional methods (Medeiros, 2002 and Medeiro, et al, 2005). Uniform distribution and fine particle size of nanosized particles of the reinforcing phase added to steel provide homogeneous dispersion of these steel carbides into the matrix, thus providing uniformity of properties and allowing compounds to be used in a variety of applications.

In the production of metal composites reinforced with carbides, powder metallurgy offers some economic and technological advantages in relation to other competing processes such as low cost of raw material processing and relatively low temperatures involved in the process. As the microstructure of the sintered steel is the result of process parameters (time, milling speed, compaction pressure, sintering time and temperature, sintering in solid or liquid state) and also the characteristics of starting powders (size and particle size distribution, compressibility and chemical purity); any changes in these parameters also affect the sintering kinetics, with wide variations in their microstructure and consequently their performance in relation to specific application. The use of the technique Powder Metallurgy (PM) in the manufacture of MMC composites is increasing.

2. Synthesis of nanostructured carbides — TaC

Refractory, ultrathin and nanostructured metal carbides (TaC, NbC, MoC and WC) have been produced at low temperature from ammonium oxalate complexes via gas-solid reactions with suitable characteristics for various purposes: high surface area (38 to $58m^2/g$), size of crystallites (18-20 nm), extremely porous particles and high acidity.

Recent discoveries aimed at facilitating the routes for obtaining these carbides with high surface area have led to a rapid growth in the application of these materials in catalysis, composite alloys (Medeiros, 2002; Oyama, 1992, and Ledoux, 1990) and metal-ceramic composites (Upadhyaya, 1998 and Matthews, 1994).

Nanosized refractory metal carbides, particularly TaC, are produced at low temperature and short reaction time through gas-solid reaction in fixed bed reactor using tris (oxalate) hydrated ammonium oxytantalate-$(NH4)_3TaO(C_2O_4)_3nH_2O$-precursor of fine grain size (ultrafine particles with high surface area). However, for successful precursor carburizing process, certain parameters such as temperature, composition of the gas mixture (H_2 and CH_4), methane concentration, reaction time, precursor cost and heating rate must be taken into account.

The system used for carrying out gas-solid reactions is composed of a horizontal tubular resistive furnace (alumina tube) with temperature controller, fixed bed reactor (slip casting), gas inlet and outlet system, valves and flow meters. The gas composition determines reduction and carburization, which is characteristic of atmosphere. The amount of CH_4 in the mixture cannot be too excessive, since much carbon would precipitate on the surface of particles, hindering the reaction and producing free carbon in the product, which hinders the diffusion process. In the case of low CH_4 concentration, carburization does not occur. The complete carbide reaction occurs at temperature below 1000°C with CH_4 concentration ranging from 3 to 5%. Under some conditions, carburization is completed within 2 hours (Medeiros, 2002).

In recent work (Lima, 2013) Fig.1, TaC synthesized revealed the formation of pure phase with characteristics different from those conventionally obtained for the product, with average crystallite size of the order of nanometers, approximately 12.05 nm, and surface area of 19 m^2/g. These characteristics allow its use not only as reinforcement for MMC composites and solid inserts but also as catalysts.

Figure 1. Electron microscopy of tantalum carbide (Lima, 2013). Magnification of 7000x.

3. Eurofer steel — Complexity microstructure and property

Eurofer steel, also called Eurofer 97 due to the year 1997 when its chemical composition was defined, is the end product of a synthesis of four steels of reduced activity that were studied in the European Union (MANET, OPTIMAX, BATMAN and OPTIFER I and II) (Lucon, *et al*, 2006). Although, a huge progress in the improvement of this steel has been developed in recent years (Lindal, 2005), studies with Euroferno with regard to optimization of the reduced activity, increased fluency and decrease in the ductile-brittle transition temperature after radioactive damage remain under constant growth, evolving thus to EUROFER 2 and 3. The aim is that EUROFER 3 becomes a residue of low level of radiation after 80 or 100 years when applied to the DEMO demonstration reactor (Daum and Fischer, 2000; Huang *et al*, 2007).

The mechanical and corrosion resistance properties of martensitic stainless steels can be seriously impaired particularly in function to the precipitation of complex phases, generally rich in chromium, operating temperature, or even during processing; thus, the thermal cycles to which these steels are submitted should be conducted under absolute control.

To provide the steel mechanical strength, hardness and toughness needed, the most common heat treatments consist of tempering followed by single or double thermal treatment. The main parameters involved in this case are: heating and cooling rate, austenitization temperature and time and tempering thermal cycles (Mariano, 2007).

In the specific case of Eurofer steel, the improvement of its chemical composition aiming at reduced activity was achieved by replacing niobium by tantalum and molybdenum by tungsten. Nb, Mo, Ni, Cu, Al and Co were restricted to ppm values. Calculations aided by computer simulation programs indicate that Eurofer is a very promising reduced-activity alloy.

The mechanical properties of ferritic-martensitic steels restricted their use to temperatures above 550°C. Subsequently, the addition of fine and homogeneously dispersed particles allowed their use at higher temperatures (650°C) to give origin to ferritic-martensitic alloys hardened by oxide dispersion (ODS) (Lindau, 2005). To date, the most promising of this series is the ODS-EUROFER alloy. This reduced-activity alloy was developed by the Karlsruhe Research Center (Forschungszentrum Karlsruhe, FZK), in cooperation with France and Russia aiming nuclear applications.

EUROFER steel (9Cr-1W) can be used in turbines for power generation, pressure vessels, nuclear reactors or applications where the material is submitted to temperatures between 250 and 550°C. One way to improve the properties of steel, so that it works at higher temperature or become more stable is to add second-phase particles into the matrix. These particles can be in the form of oxides, carbides, nitrides, or even in solid solution when certain chemicals are added to the material.

4. Metal matrix composite — MMC with ceramic reinforcement

Metal matrix composites significantly evolved from the 20's. This composite was first used in the aerospace industry. More recently, various areas of industry have increasingly studied this type of material, for example, automotive, electronic, and nuclear, among others. MMCs attract many researchers and engineers as a good alternative to develop materials with high technological potential.

Generally, the aim is to create material with the hardness of ceramics, combined with metal that has high toughness compared to ceramics. Thus, they form a family of materials called metal-ceramic materials. Nucleation and precipitation of carbides in steel to obtain good mechanical strength coupled with high toughness or the manufacture of hard metals containing high amount of carbides and / or nitrides, in which, when processed by powder metallurgy form a continuous ceramic skeleton, are classical examples of these materials (Breval, 1995).

In principle, forged steel and most conventional steels can be treated as metal matrix composites since this forged matrix has a dispersed phase. The dispersed phase may be composed of oxide, sulfide, nitride, carbide, etc. In addition, many metal alloys should be included as MMC if the microstructural definition is considered. Although the definition of a composite material is very comprehensive, there is a strong tendency to believe that the solidification direction of eutectic microstructure is within the possible definitions of MMC (Ralph, *et al*, 1997).

Different applications are found when a metal is reinforced. The reinforcement of light metals, for example, opens the possibility of their use in areas where weight reduction is a priority. The main objectives of reinforcing steel (not limited to only these) are:

- Increase in elasticity modulus and tensile strength at room temperature and above it, while maintaining minimum ductility or greater toughness;
- Increase the fluency resistance at high temperatures compared with conventional alloys;
- Increase fatigue resistance, especially at high temperatures;
- Improve the thermal-shock resistance;
- Increase corrosion resistance.

Therefore, composites are versatile in their applications because they combine the distinctive properties of each material that composes them.

Metal matrix composites can be obtained with continuous fiber reinforcement and the use of particulate reinforcements. However, particulate reinforcements have significant advantages because the cost of manufacture of such composite is reduced and conventional metallurgical processes such as powder metallurgy and casting, followed by post-processing steps such as lamination, forging and extrusion can be used. Depending on the type of particulate reinforcement, the composites obtained may have higher use temperature compared to the matrix material, improved thermal stability and wear resistance, such as Al $_{(matrix)}$ / Si $_{(fiber)}$ composites. Therefore, research efforts have been directed to obtain metal matrix composites and particulate reinforcement (Levy and Pardini, 2006).

5. High-energy milling — Reinforcement by dispersing nanoparticles into the matrix

High-energy milling is a technique fo the processing of powders in solid state involving repeated cycles of deformation, cold welding, fragmentation and cold re-welding of powder particles held in a high-energy ball mill (Suryanarayana, 1998; Suryanarayana, 2001; Koch,

1998). This method uses high-energy milling to form composite powders typically for long milling times (Gomes *et al*, 2001; Costa *el al*, 2002.).

In high-energy milling, the constant ball-powder-ball collisions result in deformations and fractures that define the dispersion of components, homogenization, phases and the final powder microstructure. The nature of these processes depends on the mechanical behavior of the powder components, their equilibrium phase and stress state during milling (Suryanarayana, 1998 and 2001). High-energy milling can be performed with three different categories of metal powders or alloys, namely: ductile-ductile components, ductile-brittle components and brittle-brittle components.

The influence of particle dispersion with the formation of a second phase, slowing the metal surface movement and its sintering, has been proposed by Kuczynsky and Lavendell, who considered that moving particles act as barriers to the surface advancement. Later, it was discussed that particles are swallowed by the surface movement exerting force and preventing its movement (Sbrockey and Johnson, 1980).

The dispersion of large amounts of ceramic particles by powder metallurgy aims to improve their tribological properties and thus their mechanical properties; therefore, 12% SiC powder with average size of 3 μm was added to 316L steel powders of 5 μm, mixing for 8 hours, uniaxially compressed at 100 MPa and sintered at temperatures of 1100°C for one hour in inert atmosphere, resulting in complete fusion of samples (Patankar, *et al*, 2000).

Another study that used 3% W TaC dispersed in the metal matrix of atomized 316L austenitic stainless steel and processed by powder metallurgy showed higher density values and a significant increase in hardness compared to material without reinforcement (Oliveira, 2008).

6. Sintering

Sintering is a non-equilibrium thermodynamic process in which a system of particles (powder or compacted aggregate) acquires a coherent solid structure by reducing the specific surface area. The result is the formation of grain boundaries and growth of necks and inter-particle bond, typically leading the system to densification and volumetric shrinkage (Gomes, 1995).

Solid-phase sintering occurs at temperature where none of the system elements reaches their melting point. This is accomplished with material transport (atomic diffusion, vapor transport, viscous flow, etc) (Costa, 2004).

In the second case, liquid-phase sintering leads to the formation of a liquid phase capable of dissolving a percentage of particles. This yields a diffusion pathway analogous to the grain boundary in the solid-phase sintering, causing a rapid initial density increase and then, dissolution of solid particles in the liquid and precipitation on the neck region occur (Gomes, 1995).

The sintering of metal matrix composites can be divided into two categories: a) solid-phase sintering (including powder metallurgy) and b) liquid-phase sintering (Kocjak, *et al*, 1993).

This is a complex step in the powder technology route and is influenced by several parameters such as: solubility, self-diffusivity and inter-diffusivity, mutual solubility and wettability, etc., which act simultaneously on the process and systems are so diverse that it has not yet been possible to develop a single sintering model capable of meeting the most varied and possible systems. Therefore, the solution is to develop a model for each system to be studied.

7. Case study: Developing a Metal Matrix Composite (MMC) — Eurofer 97 steel reinforced with tantalum carbides — TaCs

EUROFER 97 steel is a promising alloy for use in nuclear reactors or in applications where the material is submitted to working temperatures up to 550°C due to its lower strength under fluency. Factors that influence the slip of boundaries are the grain morphology, angle and speed of grain boundaries. The speed can be reduced with the presence of a dispersed phase in the material, provided it is thin and evenly distributed.

The state of Rio Grande do Norte is a major producer of refractory metals (W, Ta, Nb), ceramic minerals (diatomite, kaolin, feldspar, mica, barite, clays, etc.) and other minerals containing rare earths and semi-precious stones. However, this natural wealth that places Rio Grande do Norte among the top five producers of minerals in the country has not reversed in progress and development for the region, mainly due to lack of technology to aggregate these resources to local raw materials.

This study presents the development of a new metal matrix composite (MMC), which has the following starting materials: ferritic / martensitic EUROFER 97 stainless steel and tantalum carbides – TaC as reinforcement, one of them synthesized in laboratory (UFRN) and the other supplied by Aldrich, the first with average crystallite size of 13.78 nm and the second with crystallite size of 40.66. TaC nanometric particles were inserted into the metallic matrix through the processing steps of powder metallurgy seeking to improve the properties of the final product.

Initial sintering studies with EUROFER 97 steel reinforced by TaC nanosized particles dispersed into the matrix of MMC composites showed satisfactory values with respect to the improvement of the mechanical properties regardless of processing, as sintered materials with similar microhardness values and even greater that 100% the value of 333.2 HV for bar-shaped pure steel was obtained (Oliveira, 2013).

7.1. Experimental procedure

EUROFER steel was received in the form of bar and has ferritic / martensitic microstructure as can be seen in Figure 2, as well as the presence of grain boundaries with considerable sizes, unable to be viewed in full due to their size indicated by arrows. Microhardness measurements were made and the average value was 333.2 HV and according to the literature, the theoretical density ranges from 8.0 to 8.1 g/cm^3.

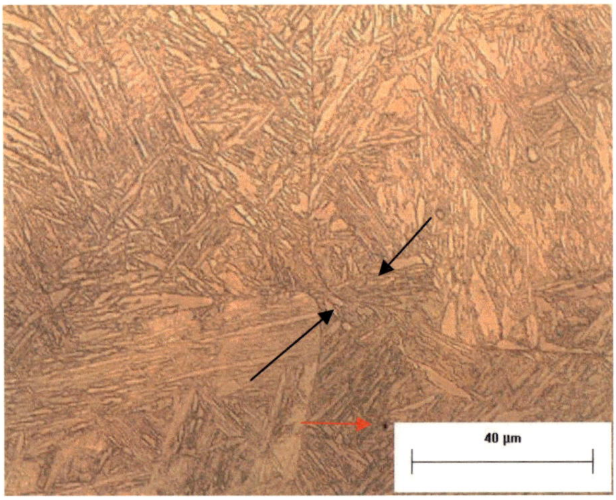

Figure 2. MO micrograph (500x) of EUROFER steel as received, attacked with 2% Vilela.

After characterization of the bar-shaped starting material, machining was performed, where chip was removed, and from this, the grinding process began. Firstly, pure steel was performed, and then steel with the addition of 3% UFRN TaC and 3% commercial TaC. The powders were milled for 5 hours in high-energy mill. Each of the resultant particulate samples were characterized by x-ray diffraction and SEM and then cold compacted under uniaxial pressure of 600MPa in a cylindrical array of 5 mm in diameter. Subsequently, the compressed powders were sintered in a vacuum oven at temperature of 1250°C isotherm for 60 minutes. Sintered samples were cooled to room temperature. The structure of sintered samples was observed by optical and scanning electron microscopy and analyzed by microhardness tests.

7.2. Results

Figure 3 shows the microstructure and the X-ray pattern of pure EUROFER steel milled for 5 hours. The electron micrograph of EUROFER 97 shows uneven particles with rough surface, which can be compared to the morphology of powder atomized with water, Fig (3a). The X-ray diffraction pattern, Figure (3b) shows only peaks associated to Fe with CCC structure, which is the same structure of the ferrite phase. This is a strong indication that the milling conditions used significantly influenced the phase transformation occurred during sintering of EUROFER steel. During milling, the starting percentage of martensite was transformed into ferrite from the effects of powder processing.

Figure (4a) shows the electron micrograph of EUROFER 97 steel composite powder with addition of elemental TaC (UFRN) after 5 hours of milling. It was observed that there is a similarity in images-SEM of pure steel powder and composite powder, Figure (4a) and Figure (4a). The presence of carbides (light spots) is noticeable, which are distributed and inserted in a dispersed form on the surface of the metal matrix (dark gray surface). Figure (4b) shows the XRD pattern of X-ray of EUROFER 97 steel composite powder with elemental TaC (UFRN)

http://dx.doi.org/10.5772/59187

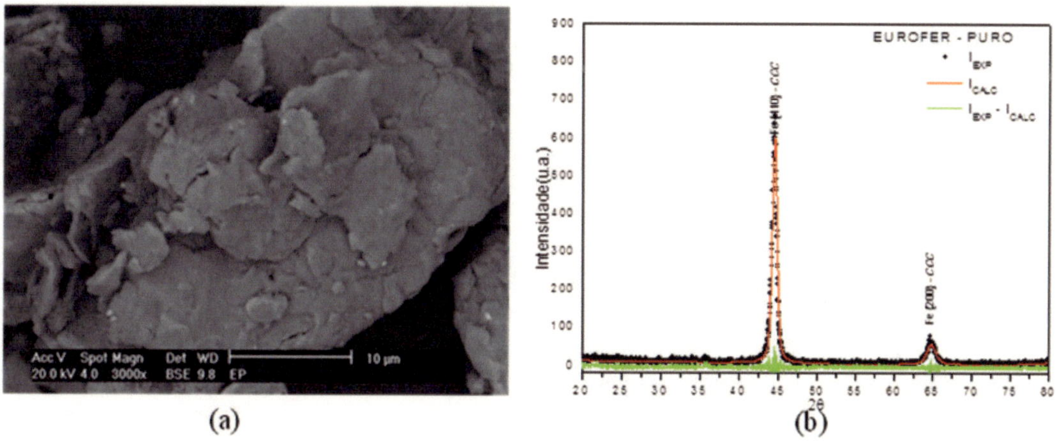

Figure 3. Pure EUROFER 97 steel powder milled for 5 hours, (a) SEM-magnification 3000x and (b) XRD.

milled for 5 hours, which presents peaks characteristics of steel and TaC, and their respective phases. The diffractogram shows only the ferrite phase, without martensite, for iron after 5 hours of milling.

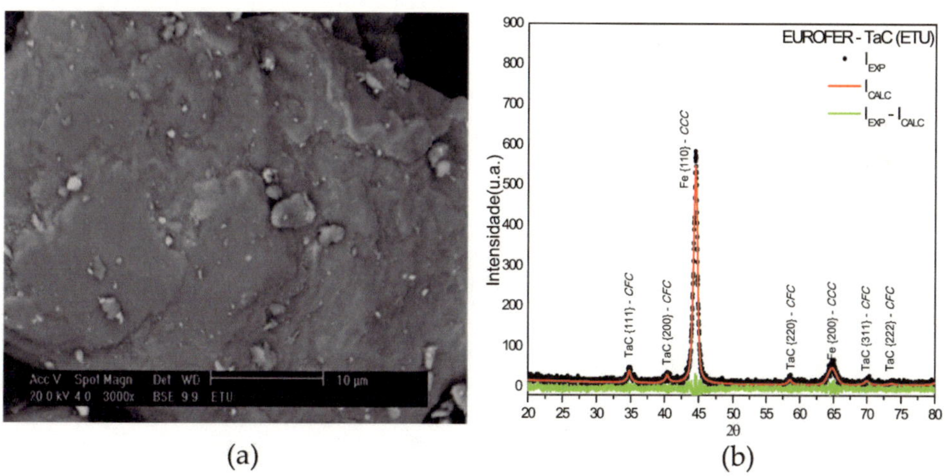

Figure 4. EUROFER 97 steel powder with TaC UFRN milled for 5 hours, (a) SEM-magnification 3000x and (b) XRD.

Figure (5a) shows the SEM image and figure (5b) shows the diffraction standard of X-ray of EUROFER 97 steel composite powder with commercial TaC milled for 5 hours in a high-energy planetary mill. Initially, the SEM microstructure shows heterogeneous particle size and shape and the non-uniform dispersion of carbides into the metal matrix compared to particulate composites of steel with UFRN TaC Figure (4a). XRD pattern shows peaks of Fe with TaC and CCC structure with greater intensity and height for carbide when compared with XRD pattern of the EUROFER 97 composite powder with UFRN TaC, Figure (4b).

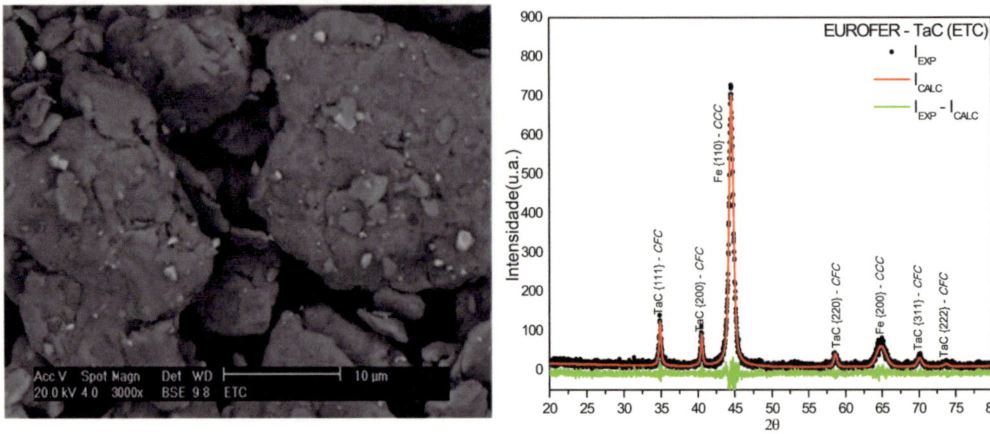

(a) (b)

Figure 5. EUROFER 97 steel powder with commercial TaC milled for 5 hours, (a) SEM-magnification 3000x and (b) XRD.

Figure 6 shows the micrographs (SEM) of samples sintered at temperature of 1250°C for 60 minutes. Figure (6a) shows sintered pure EUROFER steel with microstructure that indicates that it is in the final stage because pores are small and rounded, typical of this stage. Figures (6b) and (6c) show the dispersion of TaC particles in ETU and ETC composites, in which it is observed that TaC particles (white portion) are dispersed in the grain boundaries of the metal matrix (gray portion); in the case of ETU sample still with the presence of many pores, a large cluster of TaC particles randomly dispersed in the metal matrix between the boundaries and within the pores of the sintered sample was observed, hindering the closing of pores and the non densification of the material.

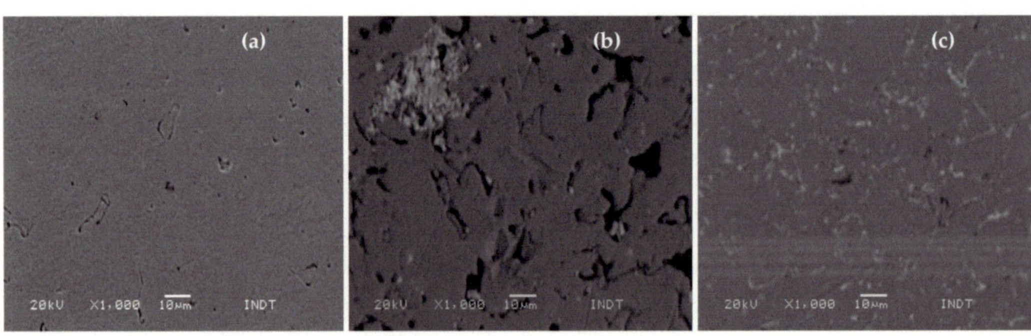

Figure 6. Micrographs (SEM-1000x) of samples sintered at 1250°C for 60 minutes.; (a) Pure EUROFER (EP), (b) EUROFER with UFRN TaC (ETU) and (c) EUROFER with commercial TaC (ETC).

Micrographs (OM) of Figure 7 show sintered pure steel (Fig. 7a), sintered steel with UFRN TaC (Fig7b) and sintered steel with commercial TaC (Fig. 7c). Regarding porosity, there are small and well rounded pores, indicating the final sintering stage. The idealized sinterization models proposed by various authors are specific to each stage. Figure 7a shows in the micrograph of sintered EUROFER pure steel sample, the size of grain boundaries and the presence of two phases, ferrite (lighter portion) and martensitic (darker portion); Figure 7b shows sintered ETU still with the presence of many pores compared to the two other sintered samples under the same conditions; however, pores in the segmented form in particles and also rounded pores, indicating sintering at stage from intermediate to the final stage. Microstructure with diffuse phases, dark and clear portions, was observed. Figure 7c shows densified samples with grain boundaries with regular shapes and small sizes; however, the presence of a single phase characteristic of ferritic phase can influence the mechanical properties of the material.

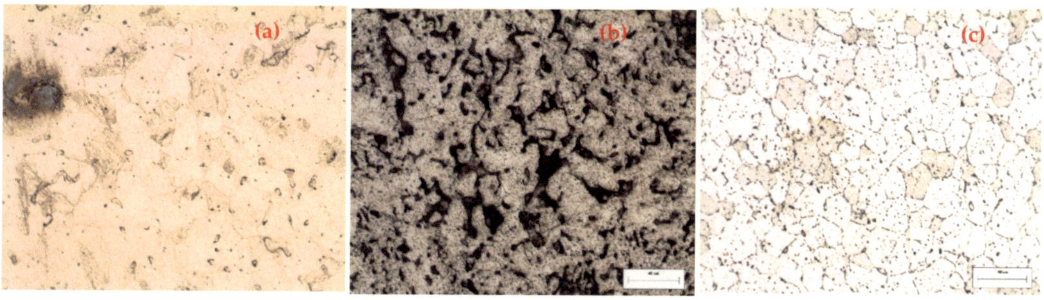

Figure 7. Micrographs (SEM-500x) of samples sintered at 1250°C for 60 minutes; (a) Pure EUROFER (EP), (b) EUROFER with UFRN TaC (ETU) and (c) EUROFER with commercial TaC (ETC).

Table 1 shows the microhardness values obtained for pure Eurofer 97 and samples added of elemental TaC sintered at temperature of 1250°C and time of 60 min.

It was observed that the ETU sample obtained the highest hardness value compared to all other samples, which can be related to the size of very fine particle, described in experimental procedure, not even when dealing with a well-densified sample.

The large difference in the microhardness values among pure steel samples can be observed, and for this, the processing of powder metallurgy (milling and sintering) was sufficient to improve the mechanical properties of steel samples with UFRN TaC compared to steel samples with commercial TaC, both at the same sintering temperature and sintering time. These differences in the microhardness values can be probably associated with the dispersal of coarse and / or fine particles in the metal matrix, and this is due to the synthesis of carbides; even with low energy input, UFRN TaC produced particles with small crystallite size and more homogeneous compared to that provided by Aldrich, or may even be related to phase transformation, i.e., in sintered samples with commercial TaC, the prevailing phase is ferrite, which has low hardness, whereas in those with UFRN TaC, the presence of other phases, i.e., the metal matrix composite remains ferritic / martensitic as the initial steel and the presence of fine particles dispersed or embedded in the metal matrix of the EUROFER steel.

Samples	Microhardness (HV)
EP	636.8 ± 35.6
ETU	700.6 ± 17.0
ETC	191.2 ± 20.3

EP= Pure EUROFER;

ETU=EUROFER + UFRN TaC;

ETC=EUROFER + commercial TaC

Table 1. Microhardness results of composites milled for 5 hours and sintered at temperature of 1250°C for 60

8. Conclusions

The process of TaC powder synthesis generated microstructure of refined grains with rounded morphology of particles, and crystallite size (13.78 nm) that may have influenced densification and also the microhardness of sintered samples.

Tantalum carbide-TaC with nanosized crystallites, strongly agglomerated, forming particles smaller than EUROFER 97 steel particles, were dispersed by milling, remaining compacted among particles that compose the metal matrix, thus obtained material with refined microstructure, smaller grain sizes and good mechanical properties.

SEM images show sintered pure steel samples and steel composites with UFRN TaC, in which microstructure typical of sintered material with initial, intermediate, final sintering stages were observed. In addition, the dispersion of tantalum carbides in the EUROFER steel metal matrix was observed.

Optical microscopy showed a change in grain size and in the microstructure of the different sintered materials. For samples with the addition of TaC, grain size decreased or remained stable compared with sintered pure steel. Sintering at temperature of 1250°C provided high densification in composite reinforced with Aldrich TaC but the microhardness result was low, which can be related to a phase change, i.e. steel became fully ferritic.

The powder metallurgy processing was sufficient to improve the mechanical properties of EUROFER steel, with small difference in relation to microhardness results between EUROFER steel and composite reinforced with UFRN TaC. In the case of composite reinforced with commercial TaC, there is a considerable difference, which can be related to the phase transformation and / or presence of coarse particles compared with UFRN TaC powder particles, which are thinner.

New EUROFER 97 steel composite with TaC was developed and characterized according to appropriate procedures for the manufacturing processes used in this work and with altered mechanical properties and future features such as: special applications in nuclear reactor, cutting materials and many others.

Author details

Leiliane Alves de Oliveira, Uílame Umbelino Gomes*, Ariadne Souza Silva and Emanuel João Batista

*Address all correspondence to: umbelino@dfte.ufrn.br

Federal University of Rio Grande do Norte, Brazil

References

[1] Breval, E. Synthesis route to metal matrix composites with specific properties: a review. Composites Engineering, v. 5, p. 1127-1133, 1995.

[2] Costa, F. A.; Ambrozio Filho, F.; Lima, N. B.; Gomes, U. U. Alves Junior, C.; Silva, A. G. P; Rodrigues, D. Synthesis of a composite powder with W-Cu nanocrystallites using tungsten grinding balls by high energy. In: XV Brazilian Congress of Materials Engineering, Nov. 9-13, Christmas. Annals, p. 2086-2092 (2002).

[3] Daum, E. Fischer, U. Long-term activation potential of the steel eurofer as structural material of a demo breeder blanket. Fusion Engineering and design.v.49-50, p. 529-533, 2000.

[4] Gomes, U. U.; Technology of Powders-Fundamentals and Applications. Christmas. UFRN University Publishing. (1995).

[5] Gomes, U. U.; Costa, F. A.; Silva, A. G. P. On sintering of W-Cu composite alloys, In:Proceedings of the 15th International Plansee Seminar, Reutte, 2001. V 1, p. 177-189.

[6] Gordo, E. ; Velasco, F. ; Atón, N. & Torralba, J.M. (2000). Wear mechanisms in high speed steel reinforced with (NbC)p and (TaC)p MMCs. Wear. Vol. 239. pp. 251-259.

[7] Huang, Q. Li, C. Li, Y Chen, M. Zhang, M. PENG, L. ZHU, Z SONG, Y. GAO, S. Progress in Development of China low Activation Martensitic Steel for Fusion Application. Journal of Nuclear Materials. v. 367-370. P. 142-146, 2007.

[8] Koch, C. C.; Intermetallic matrx composite prepared by mechanical alloying – a review, Materials Science and Engineering, v. A244, 39-(1998).

[9] Kocjak, M. J. et al. Funtamentals of metal matrix composites. Butterworth-Heinemann, Boston, n. (Ed S.Suresh, A.Mortensen, A.Needleman), 1993.

[10] Ledoux, j. m. et AL. Platimun and rhodium substitution by molybdenum and tugsten carbides for catalysis. Syntesis, activations and uses. Paris: Académie des Sciences, 310, 707-713, 1990.

[11] Levy N. F., Pardini, L. C.; Structural Composites: Science and Technology 1 ed.. Sao Paulo: Edgard Blucher, 2006.

[12] Lima, M.J.S. Synthesis and characterization of TaC and Mixed Oxide Nanostructured Tantalum and Copper from Precursor Oxaálico of Tantalum and Copper Nano etruturado from Oxalic Tantalum Precursor via Reaçõs Gas-Solid Low Temperature. 71 f.. 2013 Master Thesis-Graduate Program of Sciences and Materials Engineering, Federal University of Rio Grande do Norte, Natal-RN.

[13] Lindau, R. Möslang, A., Schirra, M., Schlossmacher, P., Klimenkov, M. Mechanical and microstructural properties of a hipped RAFM ODS steel. *Journal of Nuclear Materials*. v. 307-311, parte 1, p. 769-772, 2005.

[14] Lucon, E. Benoit, P. Jacquet, P. Diegele, E. Lässer, R. Alamo, A. Coppola, R. Gillemot, F. Jung, P. Lind, A. Messoloras, S. Novosad, P. Lindau, R. Preiningeer, D. Klimiandkou, M. Peterson, C. Rieth, M.Materna-Morris, E. Schneider, H.-C. Rensman, J.-W. Scchaaf, B. V. D. Singh, B. K. Spaetig, P. The European effort towards the development of a demo structural material: Irradiation behavior of the European reference RAFM steel EUROFER. Fusion Engineering and Design, v. 81, p. 917-923, 2006.

[15] Mariano, N.A. Pereira, V.F. Rodrigues, C.A.D. Di Lorenzo, P.L. Rollo, J.M.D.D.A. Characterization of hardenability curves and phase transformation of martensitic steels inoxidávei type FeCrNi, Revista Escola de Minas Ouro Preto, v.60 (1), p. 163-167, 2007.

[16] Martinelli A.E., D.S.A. Paulo, R.M. Nascimento, M.P. Tavora, U.U Gomes and C.A. Junior: J.Mater. Sci. Vol. 42 (2007), p. 314.

[17] Matthews, F.L.; Rawllings, R.D. Composite Materials: Engineering and Science, Chapman Hall, London (1994)

[18] Medeiros, F.F.P. Synthesis of tungsten carbide and niobium at low temperature, through gas-solid reaction in a fixed bed reactor. 2002. 145F. Thesis (Ph.D. in Chemical Engineering)-Technology Center, Department of Chemical Engineering, Graduate Program in Chemical Engineering, Federal University of Rio Grande do Norte, Natal.

[19] Medeiros, F.F.P.; Silva A.G.P.; Souza, C.P. Synthesis of niobium carbide et low its use in hardmetal. Internatinal Journal Powder Technology. Elsevier Science, 126, 155-160, 2002.

[20] Medeiros, F.F.P., C.P. Souza, C. Alves Junior, A.G.P. Silva and M.F. Pereira: Niobium carbide formation from oxalate complex with methane and hydrogen plasm reaction. In: 7th WorldCongress of Chemical Engineering. Glasgow, 2005.

[21] Oliveira, L. A.; Study of Sintering of 316L Stainless Steel reinforced with 3% tantalum carbide-TaC Dissertation in Engineering and Materials Science UFRN Christmas 2008.

[22] Oliveira, L. A.; Development and Characterization of a metal matrix composite (CMM): Eurofer97 Steel Reinforced Tantalum carbide-TaC Doctoral Thesis in Engineering and Materials Science UFRN Christmas 2013.

[23] Oyama, S. T. Preparation and Catalytic Properties of Transition Metal Carbides and Nitrides). Catalysis Today, Amsterdam: Elsevier Science, 15, 179-200,1992.

[24] Patankar, S. N., Chandrasekaran, M., TAN, M. J.; Matrix reinforcement interaction in SiC/316L stainless steel composite, *Journal of Materials Science Letters*, v.19, 613-615, (2000).

[25] Ralph, B.; Yuen, H. C.; Lee, W. B. The processing of metal matrix composites – an overview. Journal of Materials Processing Technology, v. 64, p. 339-353, 1997.

[26] Sbrockey, N. M., Johnson, D. L.; Influence of second phase particles to retard surface smoothing and sintering, *Materials Science Research*, v. 13, 177-186, (1980).

[27] Silva, A.S., Electrochemical Study of Sintered Steels reinforced with NbC and TaC for electrogeneration O2, Doctoral Thesis, UFRN, April 2005.

[28] Silva, A. S., Martinelli, A.E., Scatena Júnior, H., Echude Silva, J. H., Alves Junior, C. Electrochemical behavior of steel-FeNbC composites used in the electrogeneration of oxygen. Materials Research, june 2005, vol. 8, no. 2, p. 151-153. ISSN 1516.

[29] Silva, A.S.; Almeida, E.O.; Gomes, U.U.; Ferreira Jr, J.; Medeiros, I.A.F. Plasma Sintering of the Steel Reinforced with Nanostructured Carbides. Materials Science Forum Vols. 727-728 (2012) pp 456-461.

[30] Suryanarayana, C. Mechanical alloying. Materials Park, OH: Powder metal Technologies and applications, v. 7, 1998. 80-179 p.

[31] Suryanarayana, C. Mechanical alloying and milling. Progress in Materials Science,v. 46, p. 1-184, 2001.

[32] Upadhyaya, S.G.S.; Journal of Materials Science, v. 24, p. 3069, 1998.

Section 3

Sintering Techniques on Metals

Chapter 7

Consolidation of AISI316L Austenitic Steel — TiB2 Composites by SPS and HP-HT Technology

Iwona Sulima

Additional information is available at the end of the chapter

http://dx.doi.org/10.5772/59014

1. Introduction

Quickly developing industry and need for materials of better utility parameters contribute to development of innovative production technologies. Industry is always in search of new technologies which meet growing demands of faster better and cheaper products. Powder metallurgy is modern technology of manufacturing of ceramic, metallic and composite products. Commonly applied technologies of the powder metallurgy based on the sintering process include following stages:

i. Obtaining of the powders; the aim is obtaining of adequate chemical and phase composition of a powder of certain physical and technological properties. The powders are attained among others through: processing of natural raw materials, chemical synthesis, size reduction, segregation and other methods.

ii. Product formation; a process due to which powder is prepared giving it respective shape through pressing, drawing, casting from a slurry. During the pressing process itself, the following phenomena occur:

- particles of the powder come close at a distance allowing adhesion,
- increase of contact surface area of the powder due to relocation and their plastic deformation,
- peeling the oxide coating off through abrasion of the neighbouring particles,
- and local point sintering of the particles due to plastic deformation and increase of temperature.

iii. Sintering; main operation of the technology of the powder metallurgy, which is conducted until diffusion connection of the powder particles is established.

iv. Processing of the sinters; the final stage after which the product is ready to use.

The sintering process itself is a process depending on transforming of the powder material into new stable material with different physical properties. The characteristic feature of the technology is possibility of serial production of ready elements and semi-products of the highest quality and strictly specified composition, without impurities and faults connected with conventional technological procedures of plastic processing or foundry. Besides, parsimony in the use of materials, energy, time and automation of the process make the method very economical and prospective. Therefore, in recent years, the interest in modern sintering methods in the production processes has increased [1-4].

Sintering is the process involving many quantitative and qualitative factors. The first group of the factors affecting the sintering are values characterizing a preformed product from a power, such as: particle size of the powder, its chemical and phase composition, density of a profile, inhomogeneity of the profile. On the other hand, the second group of the factors make parameters controlled during the sintering process such as: temperature, time, pressure, sintering atmosphere. It is also to remember about so called random factors, which, in uncontrolled way, may influence the result of sintering, e.g. inhomogeneous chemical composition or inhomogeneity of mixtures [1].

From the mentioned above three groups of factors influencing properties and microstructure of the sintered materials, the most important are the following:

- Temperature of sintering, which should ensure successful finalizing consolidation processes and receiving of required utility properties. It should be selected in such a way, that during consolidation, diffusion mechanisms causing solidifying contraction dominated.
- Time of sintering is the time during which the sintering material is kept at the sintering temperature. Carefully chosen time allows for effective course of diffusion mechanisms in a material and limits also the process of grains growth.
- Pressure; application of the external load causes intensification of the consolidation processes and, among others, simplifies the elimination process of pores at the final stages of the sintering. Due to application of pressure, temperature can be lowered and time of sintering can be shorter.
- Atmosphere of sintering; during the sintering process, neutral (protective) or active atmosphere can be applied. In most cases the process is realized in the neutral atmosphere, whose main advantage is protection of the sintering materials again oxidation. This kind of atmospheres are selected in such a way, so possible reaction with a sinter is forbidden. On the other hand, the active atmospheres directly influence the sintering process causing desirable changes of the chemical composition, *eg.* prevent decomposition of the compound before finish of the sintering or reduce oxides present in the powder that impede sintering etc.
- Activators of sintering are used in order to accelerate contraction, decrease sintering temperature or formation of proper microstructure of a sinter. They influence mainly

acceleration of diffusion processes, formation of transient liquid phase and prevent grain growth [1,4-8].

Among techniques limiting grain growth and accelerating consolidation are sintering under pressure (Hot Isostatic Pressing (HIP) and High Pressure – High Temperature (HP-HT) or microwave sintering or the one taking advantage of current impulses. The main advantage of the sintering by HP-HT method is opportunity to achieve extremely high pressures together with high temperatures during the process. It is worth noting, that under influence of simultaneous operation of pressure and temperature, the process proceeds much faster (usually time does not exceed few minutes) than in the case of free sintering (which usually lasts from few to several hours). The obtained sinters are characterized by a degree of densification reaching almost 100% and isotropic properties. The use of such conditions can also reduce the diffusion of particles and prevent grain growth. The high-pressure devices are composed of hydraulic presses and specially-designed chambers that permit sintering. Presently, many solutions for such high-pressure chambers exist. Basic types of the chambers applied in industry are the following: spherical chambers of Bridgman type, 'Belt' type chambers and multi-die cubic chambers. Such construction solutions of the synthesis chamber allow for relatively large volume of reaction charge, optimal distribution of pressure and achievement of high sintering temperature. Their characteristic feature is attaining quasi-hydrostatic state of stress through stable medium transferring pressure, such as various kinds of rocks, most often. The highest possible temperature at which the sintering process can be carried by means of HP-HT method is 2000°C, or even more – depending on the duration of the sintering process [9-11]. The HP-HT method is applied to sintering large group of materials, for example: diamond [10], cubic boron nitride (cBN) [12], TiB_2 ceramics [13], gradient materials [14,15], composite materials [12,16-19] and others.

In recent years, many papers were published on new techniques of sintering, in which heating of charge is performed by impulse current source. Taylor is recognized as a progenitor of this technology [20], who in 1933 proposed a pioneering solution involving to heat up during sintering Joule heat released during current flow through the consolidating powder. This process was named resistance sintering (RS). Currently, large interest in this kind of sintering is connected with its technical and economical advantages. The resistance sintering allows to obtain dense sinters with good properties at lower temperature and short time (from few to several minutes) [21]. The RS techniques involving pulsed current include Pulse Plasma Sintering (PPS) and Spark Plasma Sintering (SPS) methods. The SPS method was developed in 1960 initially to sinter metal powders [22]. However, due to high cost of equipment and low efficiency of the sintering process, the solution has not found its application at the beginning. Only in 80's of the 20th century, a new generation of apparatus for sintering of materials was elaborated under a name Spark Plasma Sintering. The SPS method was used even then to produce modern composite and functional gradient materials [23,24]. Up till now, not all phenomena associated with pulsed current sintering have been explained. There are many theories regarding the SPS sintering. Yoshimura *et al.* proved experimentally [25] that consolidation intensifies processes caused by the pulsed current.

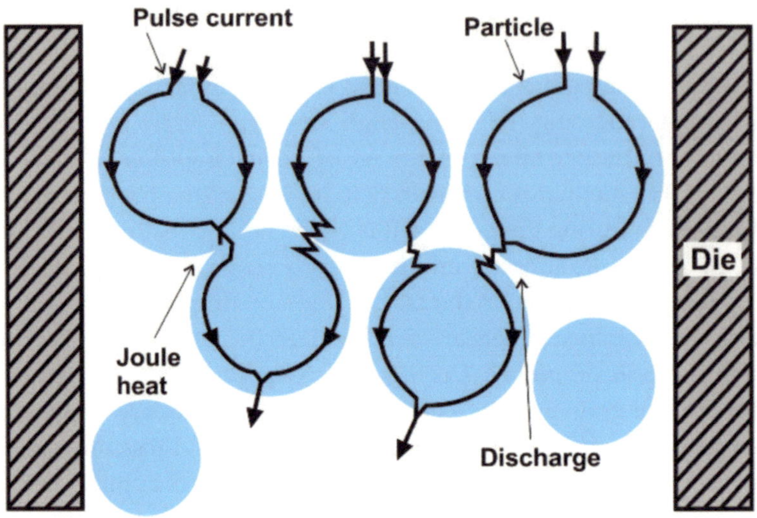

Figure 1. A schematic drawing of the pulsed current that flows through powder particles [26]

Currently accepted concept basis on the phenomenon of electric discharge. Heating up of the sintered materials is caused by pulsed current that may flow in two ways: either through graphite die and punches or through press powder particles. At the moment of current flow through powder particles, electric discharge occurs at places, where particles touch each other and at free spaces in the material (Figure 1). As a result, in these regions, a momentary increase of temperature even up to several dozen thousands degrees of Celsius is observed. The heat in the process concentrates at surface of the particles. Then, vaporizing occurs followed by cleaning, activation of the sample surface and increase of diffusion processes on the surface and grain boundaries. Next, partial melting takes place and formation of necks between particles being connected [27-29]. The pulsed current during SPS sintering is usually applied during consolidation of materials that conduct electric current. On the other hand, insulators must be consolidated with the application of the direct current. Main advantages of the SPS method are as follows:

- shortening of the time of sintering, what prohibits the grain grow of materials,
- enabling fast densification under relatively low temperatures,
- elimination of application of the sintering agents in the case of many materials,
- faster and better purifying and activation of the powder particle surfaces, what enhances sintering activity,
- consolidation of materials without the application of the initial pressing, isostatic densification and drying [26,29,30].

Thanks to SPS method, sinters of the following materials were produced in the recent years: metals and their alloys (eg. Fe, Cu, Al, Au, Ag, Ni, Cr, Ti, Mo etc.), oxide ceramics (eg. $Al_2O_3,$

MgO, ZrO_2, TiO_2, SiO_2), carbides (eg. SiC, B_4C, TiC, ZrC, WC), nitrides (eg. TiN. Si_2N4, TaN, ZrN, AlN), borides (eg. TiB_2, ZrB_2, HfB_2, VB_2), fluorides (eg. LiF, CaF_2), composite materials and intermetallic phases [24, 25, 31-35].

Currently, the TiB_2 compound belongs to the group of modern engineering ceramics. The interest in this material has been developing since the beginning of 2000. Previously, the use of TiB_2 was limited for technological reasons. TiB_2 ceramics is characterised by a unique combination of physico-chemical properties that allow it to be used under the conditions of high temperature and in corrosive environments. The most attractive properties of TiB_2 are:

- high hardness (33 GPa),
- high tensile strength (350 MPa),
- high compression strength 4900 MPa,
- the bending strength of 305-521MPa obtained for products sintered in the temperature range of 1600-1900°C,
- high Young's modulus (570 GPa) stable with increasing temperature,
- high melting point (3127°C),
- low density (4.5-4.62 g/cm^2).

Titanium diboride is also characterised by a very good resistance to oxidation, chemical and structural stability at high temperatures, thermal shock resistance and abrasion resistance [36-38]. Currently, TiB_2 ceramics is an attractive material for specific applications in the aviation, automotive, defence, and aerospace industries, including the production of armours for land vehicles, ships and planes, aerospace parts and parts operating at high temperatures, characterised by very good abrasion resistance. Restrictions on more extensive use of TiB_2 ceramics are mainly due to difficulties associated with the sintering process and obtaining in this way a pure ceramic material of suitable density. They include, among others:

- a very high melting point,
- low coefficient of diffusion along the grain boundaries, requiring a long time of sintering,
- presence of oxide layers of TiO_2 and B_2O_3 adversely affecting the compaction process and reaction with metal matrix [39,40].

Despite these difficulties of purely technological character, recent years have brought an increased interest in titanium diboride, with focus on its use as a phase reinforcing the metal matrix composites. Technical literature [41-46] presents studies that relate to the use of TiB_2 ceramics as a reinforcing phase of composites based on, among others, iron, aluminium, copper, titanium, or cobalt, and on their respective alloys. Authors of the research works focus their attention mainly on studies of the effect of the amount of the reinforcing phase on the properties and microstructure of these composites and on the development of best technology for their manufacture. In the composites reinforced with particles of TiB_2, an important issue is to choose the right size of the reinforcing phase and at the same time its optimum percent content in the composite material. Depending on the volume fraction and particle size of each material, sintered products with highly varied microstructure and properties can be obtained. Studies described in [48,49] show the impact of TiB_2 ceramics on microstructure and tribological as well as mechanical properties of austenitic stainless steel sintered by HIP. Nahme et al. [47] studied the mechanical properties of AISI 316L stainless steel reinforced with 15 vol% of TiB_2, including its behaviour at elevated temperatures. An improvement was obtained in the Young's modulus (218 GPa), tensile strength (885 MPa) and compression strength (1800 MPa) of the sintered materials. As proved by various research works, the deformation of sintered AISI 316L stainless steel reinforced with 15 vol% of TiB_2 was significantly reduced from 45% to 6%. Microstructural observations revealed a uniform distribution of TiB_2 ceramics in the examined material and presence of phases rich in Cr/Mo in places where the grains of this ceramics appeared. Tjong et al. [48] examined the properties of AISI 304 austenitic stainless steel reinforced with varying amounts of TiB_2 ceramics. It has been shown that increasing the content of the reinforcing TiB_2 phase improved both hardness and tensile strength, but at the expense of reduced ductility. Based on the tribological tests carried out, a dramatic improvement of abrasion resistance has been obtained with increased volume fraction of the ceramic phase.

The aim of the presented study was to analysis of the effect of sintering techniques on the physical, mechanical and tribological properties of 316L austenitic steel-TiB_2 composites.

2. Materials and research methodology

2.1. Materials selected for sintering composites

The TiB_2 powder (2.5-3.5μm in average size, 99.9 wt.% in purity) and commercial AISI 316L austenitic stainless steel (25μm in average size, Hoganas) were used in the present study. The stainless-steel powder have the chemical composition as follows: 17.20 wt.% Cr, 12.32 wt.% Ni, 2.02 wt.% Mo, 0.43 wt.% Mn, 0.89 wt.% Si, 0.03 wt.% S, 0.028 wt.% P, 0.03 wt.% C and balance of Fe. Figure 2 presents the morphology of the selected powders.

The raw powders were mixed in the special, closed container using the mixer of TURBULA type. The powders were mixed for 8 hours. The initial phase composition of mixtures for the samples preparation were as follow:

- Steel AISI 316L+2 vol% TiB_2
- Steel AISI 316L+4 vol% TiB_2
- Steel AISI 316L+6 vol% TiB_2
- Steel AISI 316L+8 vol% TiB_2

For composition analysis, the AISI 316L austenitic stainless steel was used.

Figure 2. Morphology of powders of: a) titanium diboride, b) AISI 316L austenitic stainless steel and c) mixtures of steel with 8 vol% TiB_2 used in current study.

2.2. Fabrication of composites

Consolidation of composites reinforced with various volume fractions of TiB_2 were performed by High Pressure-High Temperature (HP-HT) and Spark Plasma Sintering (SPS). First, for the densification of composites the high pressure–high temperature (HP–HT) Bridgman type apparatus was used. Figure 3 presents the scheme of Bridgman-type HP-HT apparatus. The samples were sintered at temperature of 1000°C and 1300°C pressure of 5 ± 0.2 GPa for 60 seconds. In the case this methods, the samples had a size: 15 mm in diameter, 5 mm in high.

Next, the SPS process was carried out using the HPD 5 FCT System GmbH furnace. Materials are compacted in a graphite die using a maximum pressure of 35 MPa at vacuum. Maximum pressure was obtained after duration of 10 minutes. Vacuum and pressing time of 10 minutes was to vent the mixture. After that, the SPS furnace chamber introduced argon, which acted as a protective gas and the sintering process was carried out. The heating rate and holding time were 200°C*min^{-1} and 30 minutes, respectively. In the case this methods, the samples had a size: 20 mm in diameter, 7 mm in high. Figure 4 shows the schematic of Spark Plasma Sintering (SPS).

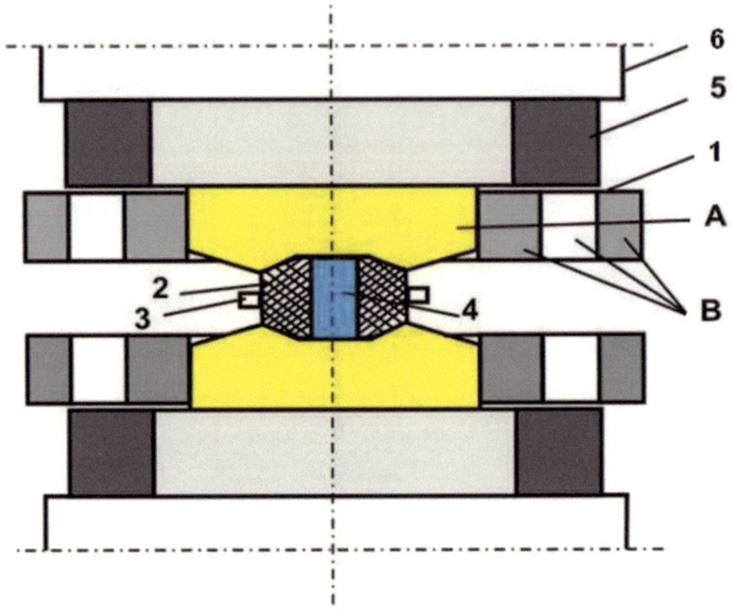

Figure 3. Scheme of Bridgman-type, toroidal HP-HT apparatus: 1 – anvil (A – central part made of sintered carbides, B – supporting steel rings), 2– pyrophyllite container, 3 – pyrophyllite gasket, 4 – material for sintering, 5 – punch, 6 – supporting plate [11].

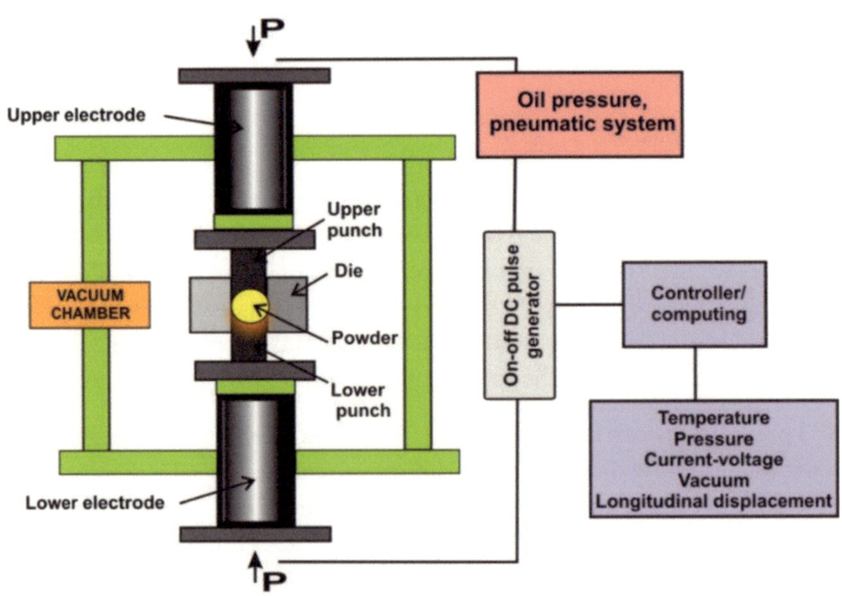

Figure 4. Spark plasma sintering device scheme.

2.3. Testing methods

Density was determined by weighing in air and water using Archimedes method. Uncertainty of measurements was 0.02 g/cm³. Young's modulus of the composites were measured basing on the velocity of the ultrasonic waves transition through the sample using ultrasonic flaw detector Panametrics Epoch III. The velocities of transverse and longitudinal waves were determined as a ratio of sample thickness and relevant transition time. The accuracy of calculated Young's modulus is estimated at 2 %. Calculations were made using the following formula:

$$E=\rho C_T^2\frac{3C_L^2 - 4C_T^2}{C_L^2 - C_T^2}$$

where: E-Young's modulus, C_L – velocity of the longitudinal wave, CT-velocity of the transversal wave,ϱ-density of the material.

A JEOL JSM 6610LV scanning electron microscope with Energy Dispersive Spectrometry (EDS) and Hitachi SU-70 scanning electron microscope with Wavelength Dispersive Spectroscopy (WDS) were used for microscopy studies of composites. Also, the phase compositions of selected samples were analysed by X-ray diffraction (XRD) using Cu K radiation with a scintillation detector (Empyrean; PANalytical).

The Vickers microhardness (HV0.3) of the composite samples was measured at FM-7 microhardness tester. The applied load for non-graded materials was 2.94 N. Standard deviations of HV 0.3 values were no more than 4 % of the average values. The compression test was carried out at INSTRON TT-DM machine at strain rate of about of 10^{-3} s.

Tribological tests were performed using a ball-on-disc wear machine UMT-2T (producer CETR, USA). The experimental procedure followed the ISO 20808:2004(E) [49,50]. For ball-on-disk method the sliding contact is brought by pushing a ball on a rotating disc specimen under a constant load. The loading mechanism applies a controlled load F_n to the ball holder. The friction force was measured continuously during the test using the extensometer. For each test a new ball is used. Specimens were washed in high purity acetone and dried. After the ball and sample were mounted, materials are washed in ethyl alcohol and then dried. The wear test conditions were:

- ball made of $Al_2O_{3,}$ diameter of 3.175 mm,
- friction track diameter: 4 mm,
- sliding speed: 0.1 m/s,
- total sliding distance: 200 m,
- test duration: 2000s,
- load applied: 4 N,
- room temperature.

The values of friction coefficient were calculated from the following equation:

$$\mu = \frac{F_f}{F_n L}$$

where F_f is the measured friction force, and F_n is the applied normal force, L – sliding distance [m].

Following the wear test, the specific wear rate was calculated. For the wear track on the disc specimen, the cross-sectional profile of the wear track at four places at intervals of 90° using a contact stylus profilometer was measured with accuracy of measurement in the vertical axis of 0.01 μm, in the horizontal axis of 0.1 μm. The cross-sectional area of the wear track was calculated using a custom-developed software. Specific wear rate according to wear volume was calculated by means of equation:

$$W_{V(disc)} = \frac{V_{disc}}{F_n L}$$

where: $W_{V(disc)}$ – specific wear rate of disc [mm^3/Nm]; V_{disc} – wear volume of disc specimen [mm^3]; F_n – applied load [N]; L-sliding distance [m].

The wear volume of disc specimen was calculated from following Equation:

$$V_{disc} = \frac{}{2} R(S_1 + S_2 + S_3 + S_4)$$

where: R –radius of wear track [mm]; S_1 to S_4 – cross-sectional areas at four places on the wear track circle [m^2].

The cross-sectional microstructure of worn surface was observed using a scanning electron microscope (SEM) JEOL JSM 6610LV.

3. Results

Figure 5 shows the results of measurements of the apparent density of the composite materials sintered by HP-HT and SPS. From these results (Figure 5a) it follows that all the materials sintered by HP-HT have reached a very high level of density, despite the short sintering time (60 sec). This was due to the combined effect of high pressure and high temperature during the HP-HT sintering process. The use of such sintering conditions allows restricting the diffusion and prevents growth of grains in the sintered materials. For composites sintered at 1300°C, the apparent density (Table 1) was in the range of 99-100% of the theoretical density. The apparent density was found to be decreasing with the increasing amount of the reinforcing TiB_2 phase in the composite matrix. This was the result of a much lower density of titanium diboride (4.52 g/cm^2 [55]) compared with the density of AISI 316L austenitic stainless steel (8.00

g/cm² [56]). The resulting sintered composites had low porosity. The porosity of the composites with various content of the TiB_2 ceramics was in the range of 0.004-0.017% (Table 1).

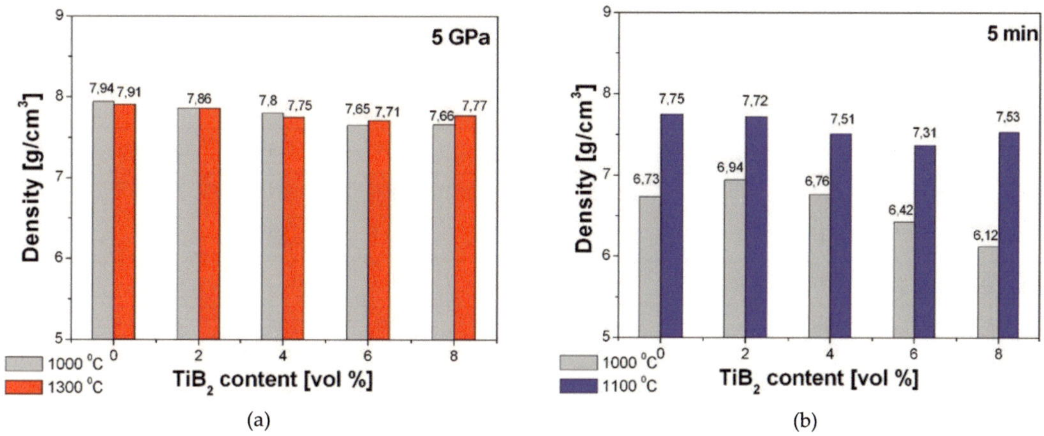

Figure 5. Effect of TiB_2 particle content and sintering methods: a) HP-HT [51,52] and b) SPS [53] on relative density of composites.

In contrast, analysing the results (Figure 5.b) of the apparent density measurements for the composites sintered by SPS, a strong correlation was noticed between the apparent density of the tested materials and the sintering temperature (Table 2). The apparent density was significantly improved with the increase of sintering temperature. The best density in the range of 97-99% of the theoretical density was obtained for the sintering temperature of 1100°C, while sinters produced at a temperature lower than 1000°C had the density of 82-88%. The temperature of 1000°C was not sufficient to obtain the required degree of consolidation in the austenitic steel-based composites reinforced with TiB_2 ceramics. Careful examination of the results has showed that the process of composite consolidation was most intense during the first few minutes of the SPS, producing the required degree of density (Figure 5b). Materials manufactured at 1000°C were characterised by high porosity in the range of 10.5-18.5% (Table 2). Increasing the sintering temperature to 1100°C has improved this parameter, and composite porosity was found to be at a level of 0.03-2.3%.

Sintered materials	Sintering conditions (HP-HT) T [°C]	p [GPa]	Relative density $\varrho 0$ [g/cm³]	$\frac{\varrho_0}{\varrho_{teor}}$ [%]	Porosity [%]	([-]	Young's modulus E [GPa]	$\frac{E}{E_{teor}}$ [%]
steel AISI 316L	1000	5	7.94	100	0.005	0.30	190	91
	1300		7.91	100	0.004	0.31	192	94
steel AISI 316L+ 2vol% TiB_2	1000	5	7.86	100	0.006	0.30	197	91

Sintered materials	Sintering conditions (HP-HT)		Relative density	$\frac{\varrho_0}{\varrho_{teor}}$	Porosity	(	Young's modulus	$\frac{E}{E_{teor}}$
	T [°C]	p [GPa]	ϱ0 [g/cm³]	[%]	[%]	[-]	E [GPa]	[%]
	1300		7.86	100	0.004	0.30	205	94
steel AISI 316L+ 4vol% TiB_2	1000	5	7.8	100	0.005	0.30	202	90
	1300		7.75	99	0.010	0.29	209	93
steel AISI 316L+ 6vol% TiB_2	1000	5	7.65	99	0.009	0.29	208	91
	1300		7.71	99	0.011	0.29	215	94
steel AISI 316L+ 8vol% TiB_2	1000	5	7.66	100	0.007	0.29	207	88
	1300		7.77	100	0.004	0.29	216	93

Table 1. The selected physical properties of composites with the various content of TiB_2 sintered by HP-HT method [51,52].

Sintered materials	Sintering conditions (SPS)		Relative density	$\frac{\varrho_0}{\varrho_{teor}}$	Porosity	(	Young's modulus	$\frac{E}{E_{teor}}$
	T [°C]	time [min]	ϱ0 [g/cm³] [54]	[%]	[%]	[-]	E [GPa]	[%]
AISI 316L	1000	5	6.73	83	12.32	0.26	165	79
	1100		7.75	97	1.12	0.27	196	94
steel AISI316L+ 2vol% TiB_2	1000	5	6.94	88	10.55	0.29	138	64
	1100		7.72	98	0.03	0.27	207	96
steel AISI316L+ 4vol% TiB_2	1000	5	6.76	86	10.89	0.29	130	58
	1100		7.51	96	0.08	0.28	202	90
steel AISI316L+ 6vol%. TiB_2	1000	5	6.42	83	16.29	0.26	112	49
	1100		7.49	96	2.10	0.28	203	89
steel AISI316L+ 8vol% TiB_2	1000	5	6.29	82	18.27	0.27	166	67
	1100		7.53	98	0.034	0.28	215	91

Table 2. The selected physical properties of composites with the various content of TiB_2 sintered by SPS method.

Figure 6a shows the results of studies of the Young's modulus of materials sintered by HP-HT. The results demonstrate an obvious improvement in this parameter when the content of the reinforcing TiB_2 phase is increasing in the composite matrix. The highest values of Young's

http://dx.doi.org/10.5772/59014

modulus were obtained in the composites reinforced with 8 vol% of TiB_2 (Figure 6a). For comparison, Young's modulus was changing with the increasing content of the reinforcing phase from 197 GPa to 207 GPa for the sintering temperature of 1000°C, and from 192 GPa to 216 GPa for the sintering temperature of 1300°C, thus proving that the temperature of sintering also exerts an effect on this parameter (Table 1 and Figure 6a). With the application of temperature raised to 1300°C, higher values of the Young's modulus were obtained. Also in composites sintered by SPS, the temperature of sintering had a significant effect on the Young's modulus (Figure 6b and Table 2). At 1000°C, low values of the Young's modulus and a large scatter in the results of measurements (in the range of 112-176 GPa) were obtained. This might be due to the high porosity of materials sintered under such conditions. Raising the temperature to 1100°C considerably improved the values of Young's modulus (202-215 GPa). In general, it has been found that the temperature of 1000°C is not preferred in the case of the SPS method, since it does not ensure the required high physical properties of the tested materials.

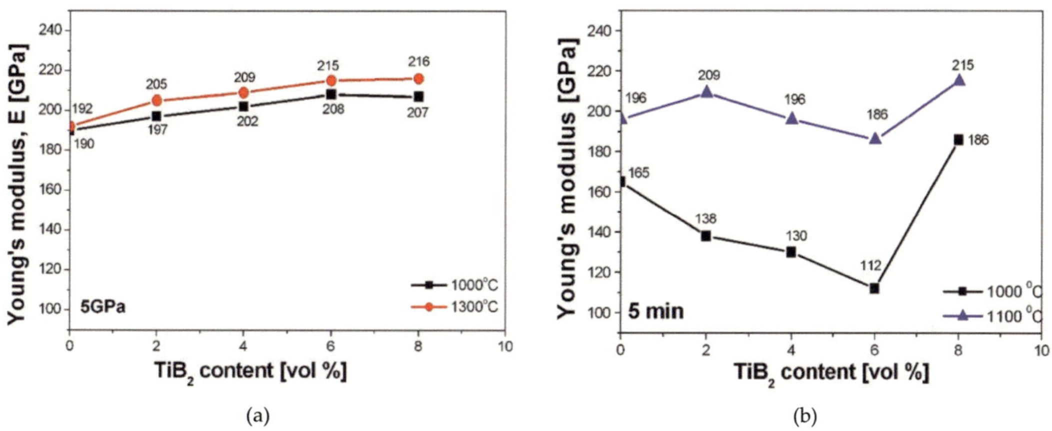

Figure 6. Effect of TiB_2 particle content in matrix and sintering methods: a) HP-HT [51,52] and b) SPS on Young's modulus of composites.

Figures 7 and 8 show examples of the microstructure produced in AISI 316L austenitic stainless steel and in the examined composites made by HP-HT and SPS. In all the composite materials, microstructural observations revealed a homogeneous distribution of the reinforcing TiB_2 phase in the composite matrix. The best effect of the uniform distribution of TiB_2 was obtained in the composites containing 6 vol% and 8 vol% of this compound (Figure 7de, 8de). In all the tested composites, the reinforcing TiB_2 phase had the tendency to settle along the matrix grain boundaries.

Microscopic observations have proved that composites obtained by the HP-HT technique were nearly free from porosity (Figure 7b-e). It was the result of a very high degree of consolidation during the sintering process and a very high level of the obtained density. Figure 7a shows the microstructure of AISI 316L austenitic stainless steel after sintering by HP-HT. A characteristic feature of this microstructure is the presence of fine and irregular austenitic grains of a few μm. A similar structure of the steel matrix was also observed in the composites with 2 vol% of TiB_2 (Figure 7b). On the other hand, the steel matrix in the composites with higher volume fraction of TiB_2 was characterised by a different type of morphology including oval-shaped grains with smooth boundaries. Microstructural analysis made by EDS and WDS (Figure 10,11) and X-ray studies (Figure 9a) confirmed the presence of titanium diboride in the form of dark precipitates settled at grain boundaries. All the investigated microstructures were observed to show the local presence of precipitates containing large amounts of nickel (Figure 11).

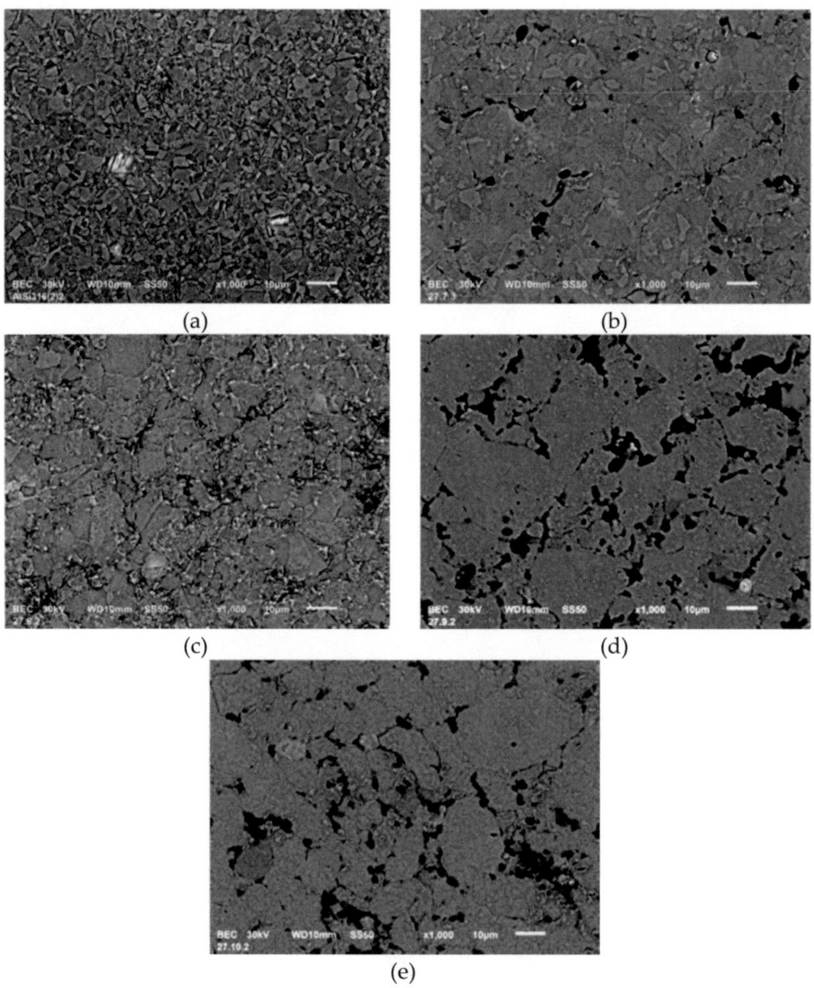

Figure 7. Selected micrographs: a) AISI 316L austenitic stainless steel and composite with: b) 2 vol.% of TiB_2 c) 4 vol% of TiB_2 d) 6 vol% of TiB_2 and e) 8 vol% of TiB_2 (HP-HT method, sintered at 1300°C and 5 GPa).

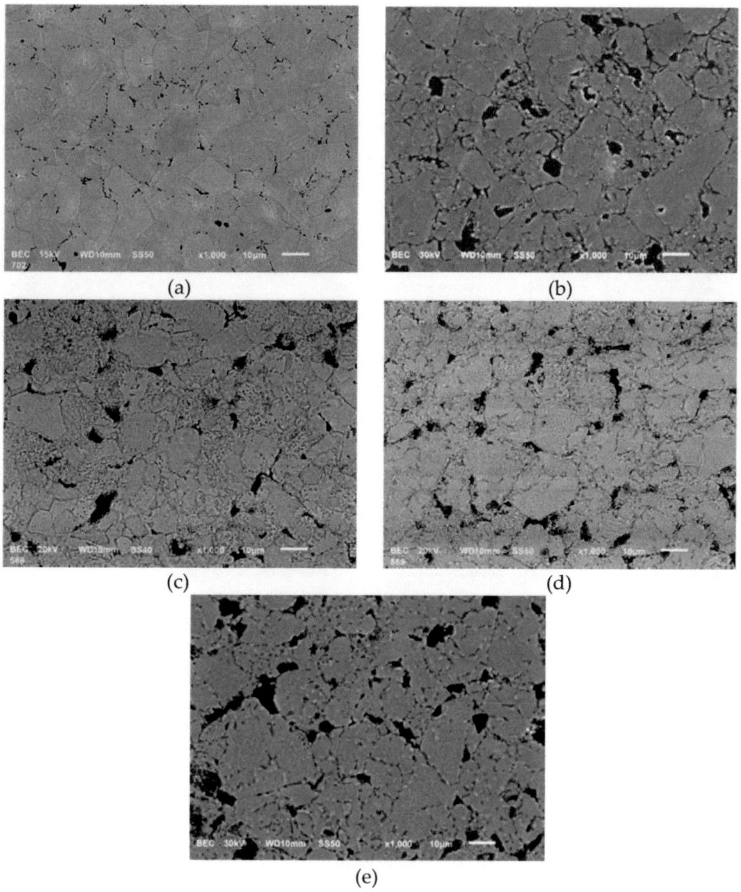

Figure 8. Selected micrographs: a) AISI 316L austenitic stainless steel and composite with: b) 2 vol% of TiB_2 c) 4 vol% of TiB_2 d) 6 vol% of TiB_2 and e) 8 vol% of TiB_2 (SPS method, sintered at 1100°C and 5 min).

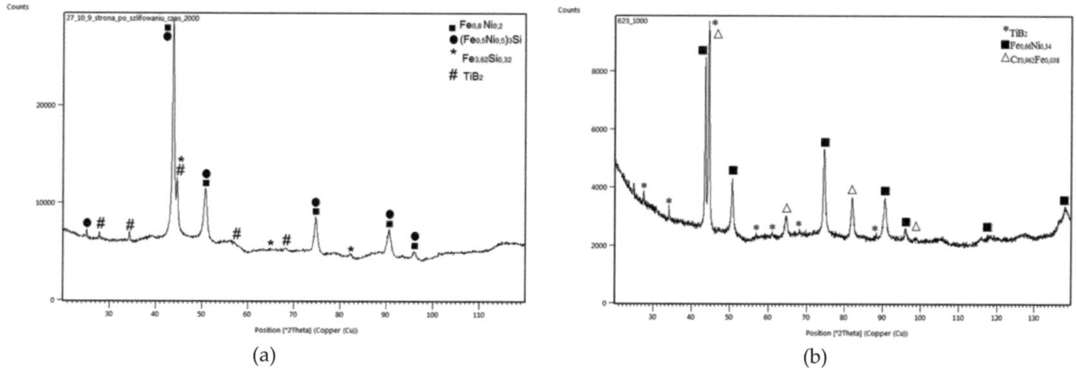

Figure 9. XRD patterns of the composites with 8 vol% TiB_2 particles obtained by HP-HT (1300°C) and SPS method (1100°C).

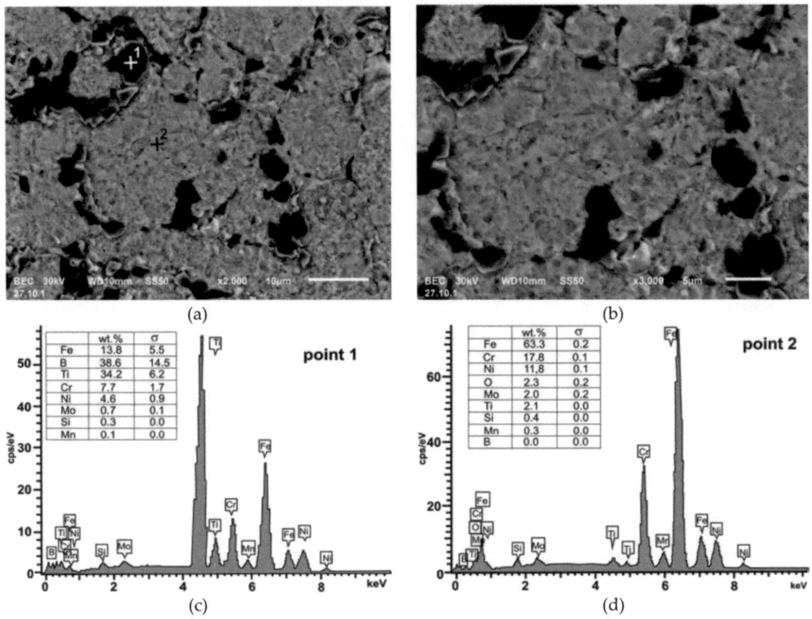

Figure 10. (a, b) The microstructure of composites with 8 vol% of TiB_2 (HP-HT method, sintered at 1300°C, 5 GPa) and (c, d) corresponding point analysis (EDS).

In all the composites made by SPS, the microstructure of the steel matrix has prevailed, i.e. the one characterised by the presence of particles with regular oval shapes, along which the reinforcing TiB_2 phase was distributed (Figure 8b-d). Sinters obtained at a temperature of 1000°C showed some voids at the grain boundaries. In contrast, the composites sintered at 1100°C were nearly completely free from porosity. This demonstrates fully effective optimisation of the SPS sintering process and selection of the best sintering conditions. Differences were revealed in the microstructure of materials sintered by SPS as compared to the materials sintered by high-pressure method. Chemical analysis (EDS and WDS) of composites sintered by SPS disclosed in the microstructure the presence of numerous precipitates containing mainly chromium (Figure 12,13), while X-ray analysis (Figure 9b) further confirmed the presence of titanium diboride and chromium-containing phases. Additionally, in the immediate vicinity of TiB_2 ceramics (Figure 13, the formation of large precipitates with characteristic band substructure was observed (Figure 13). The examined results of microstructural studies clearly show that both the sintering method and the sintering conditions have a significant impact on changes in the microstructure of the tested composites. Application of the SPS method promotes the formation of new phases. Some attention deserves the fact that the SPS sintering process uses pulsed current for heating of the sintered materials. At the time of current flow through the grains, spark discharges take place at the grain contact points and in the void areas of the sinter. As a result of these effects, a temporary temperature increase (up to several thousand degrees Celsius) can be expected in the near-surface layer of particles. It is followed by evaporation, cleaning and activation of the sample surface and by the increased rate of diffusion on the surface and along the grain boundaries [29,30]. A combination of these mechanisms operating in the SPS accelerates the consolidation process and brings changes to the microstructure of the tested composites.

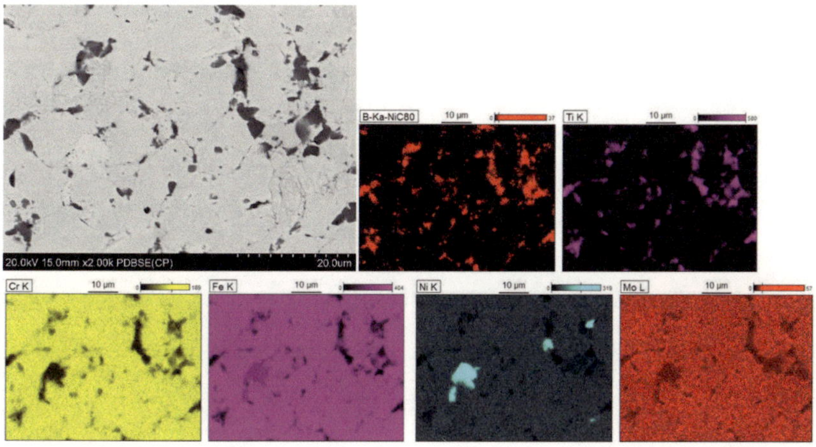

Figure 11. The microstructure of composites with 8 vol% of TiB_2 (HP-HT method, sintered at 1300°C, 5 GPa) with corresponding area analysis (WDS).

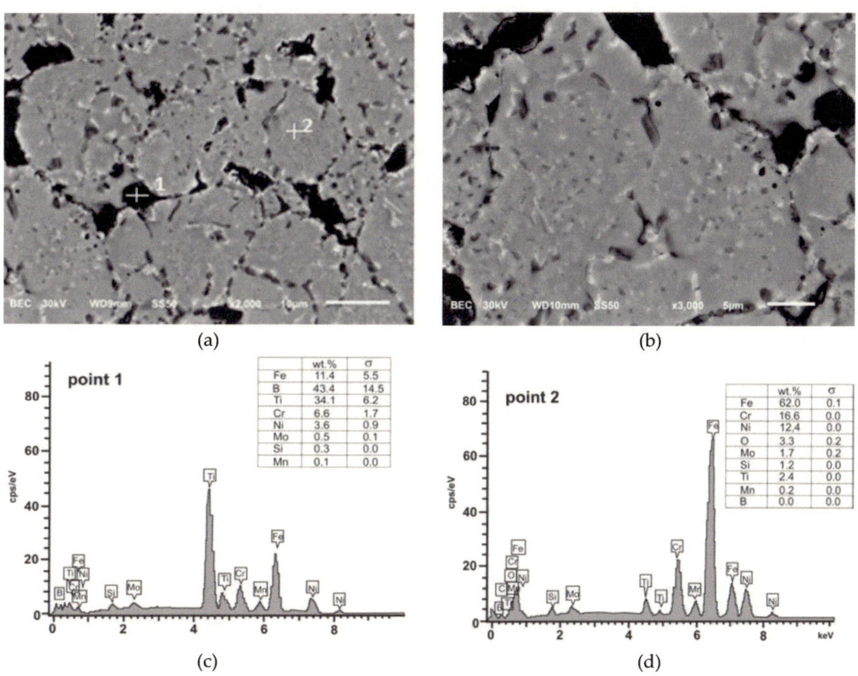

	wt.%	σ
Fe	11.4	5.5
B	43.4	14.5
Ti	34.1	6.2
Cr	6.6	1.7
Ni	3.6	0.9
Mo	0.5	0.1
Si	0.3	0.0
Mn	0.1	0.0

	wt.%	σ
Fe	62.0	0.1
Cr	16.6	0.0
Ni	12.4	0.0
O	3.3	0.2
Mo	1.7	0.2
Si	1.2	0.0
Ti	2.4	0.0
Mn	0.2	0.0
B	0.0	0.0

Figure 12. (a, b) The microstructure of composites with 8 vol% of TiB_2 (SPS method, sintered at 1100°C, 5 min) and (c, d) corresponding point analysis (EDS).

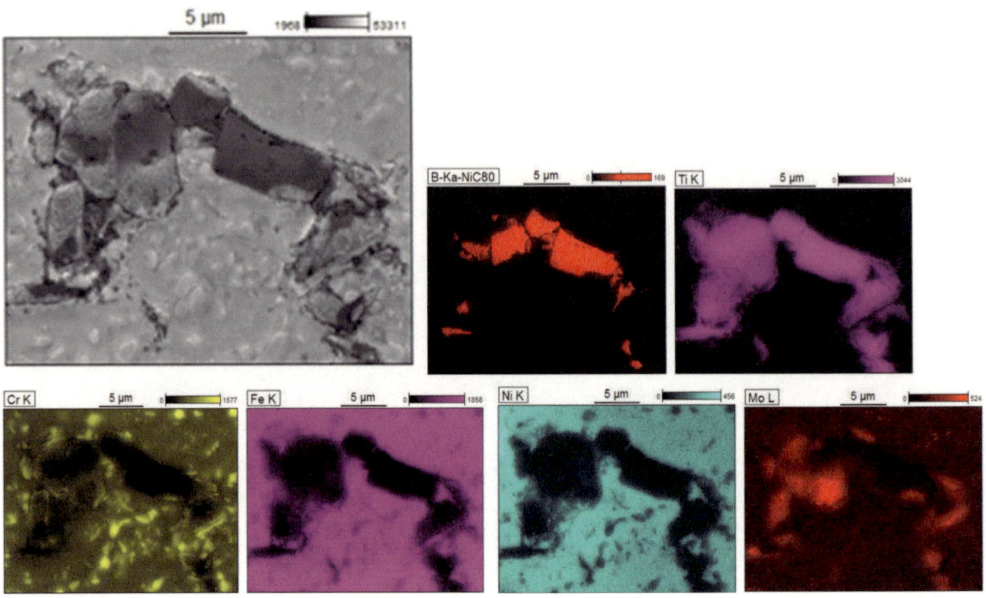

Figure 13. The microstructure of composites with 8 vol% of TiB_2 (SPS method, sintered at 1100°C, 5 min) with corresponding area analysis (WDS).

Figure 14 shows the results of microhardness measurements. In the case of the HP-HT method (Figure 14a) microhardness was found to be essentially dependent on the sintering temperature. In materials sintered at 1000°C, the microhardness higher by about 20-40%, compared to the microhardness of the composites sintered at 1300°C, was obtained. For example, in the composite with 8 vol% of TiB_2, the microhardness was 368 HV0.3 and 282 HV0.3 for the temperature of 1000°C and 1300°C, respectively. All the composites sintered by SPS (Figure 14b) suffered a definite decrease of microhardness. For materials sintered at 1000°C and 1100°C, a very low level of microhardness was obtained. It was comprised in the range of 170-265 HV0.3 and depended on the TiB_2 content and sintering temperature.

The results of compression tests are shown in Figures 15-17. Careful analysis of the data obtained for composites sintered by HP-HT proves that the introduction of TiB_2 ceramics to the steel matrix significantly improves the mechanical properties of the tested composite materials (Figure 16), compared with the steel without reinforcement (Figure 15a). High content of the TiB_2 reinforcement in the composite matrix improves also the compression strength of this material. The compression strength of the steel without reinforcement was at a level of 580-730 MPa, depending on the sintering temperature (Figure 15a), while even the addition of TiB_2 as small as 2 vol% introduced to the steel matrix raised the strength to 1000-1160 MPa (Figure 16a). For composites with 8 vol% of TiB_2, the compression strength in the range of 1200-1300 MPa was obtained (Figure 16d). The compression curves of the composites sintered at a temperature lower than 1000°C (HP-HT method) were characterised by a large range of elastic deformation and more narrow range of plastic deformation. In contrast, in the composites obtained at 1300°C, the analysis of compression curves revealed a

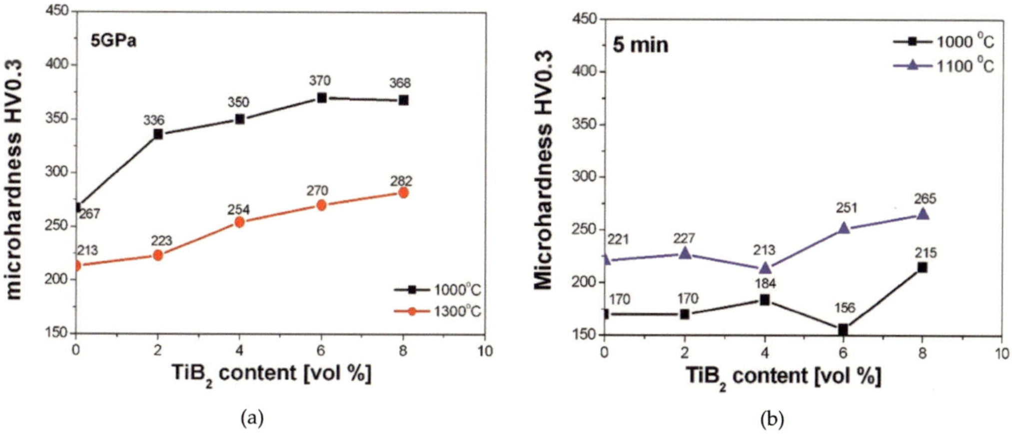

Figure 14. Effect of TiB_2 particle content in matrix and sintering methods: a) HP-HT [51,52] and b) SPS [53] on microhardness of composites.

large range of plastic deformation accompanied by a double reduction in the range of elastic deformation. All the composites sintered by SPS showed very similar characteristics of the stress-strain curve with a large range of plastic deformation (Figure 17). In general, the mechanical properties of composites sintered by SPS were inferior to the properties of composites sintered by high-pressure technique. The compression strength of the AISI 316L austenitic stainless steel without reinforcement was in the range of 520-780 MPa, depending on the sintering temperature (Figure15b). Composites sintered at 1100°C offered higher compression strength comprised in the range of 930-1040 MPa, depending on the content of the reinforcing phase in matrix (Figure 17).

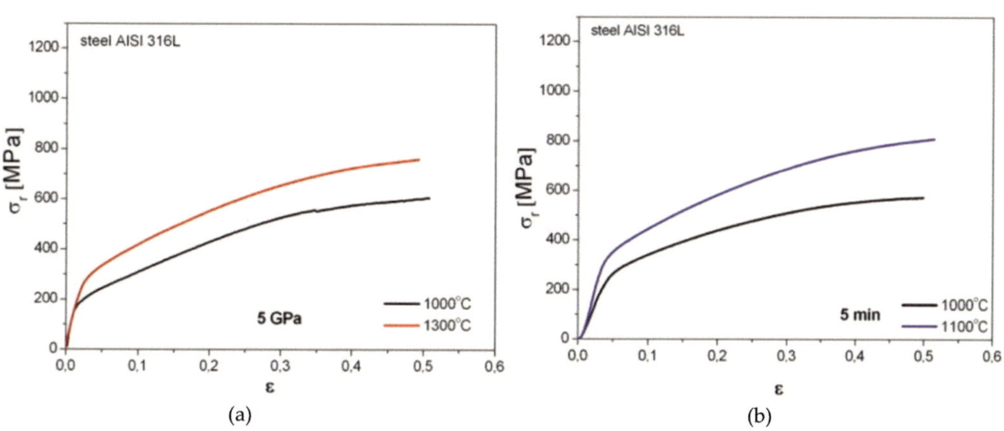

Figure 15. Compression strength of AISI 316L austenitic stainless steel sintered by: a) HP-HT and b) SPS methods.

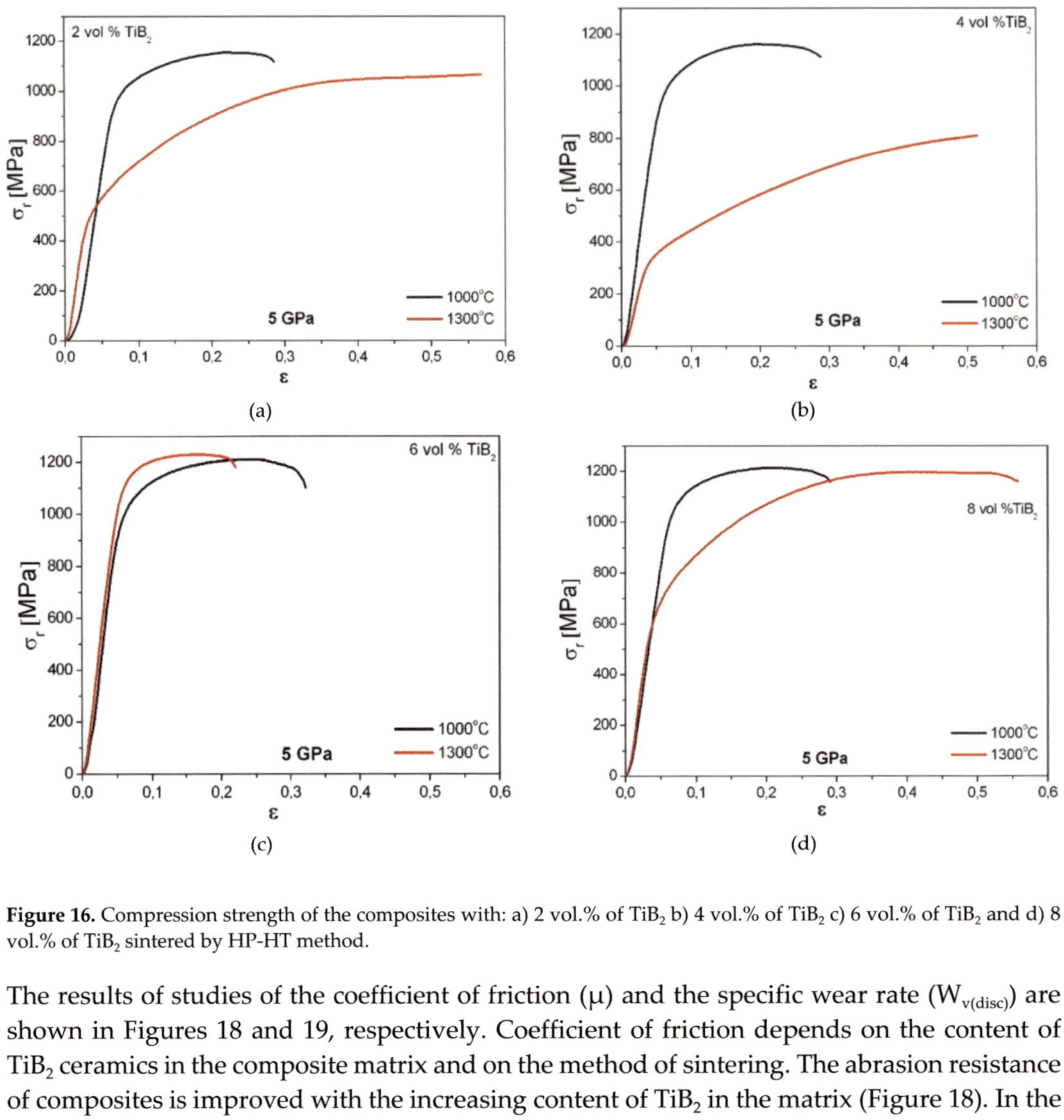

Figure 16. Compression strength of the composites with: a) 2 vol.% of TiB_2 b) 4 vol.% of TiB_2 c) 6 vol.% of TiB_2 and d) 8 vol.% of TiB_2 sintered by HP-HT method.

The results of studies of the coefficient of friction (μ) and the specific wear rate ($W_{v(disc)}$) are shown in Figures 18 and 19, respectively. Coefficient of friction depends on the content of TiB_2 ceramics in the composite matrix and on the method of sintering. The abrasion resistance of composites is improved with the increasing content of TiB_2 in the matrix (Figure 18). In the case of HP-HT sintering, the lowest coefficient of friction was obtained in composites containing 8 vol% of TiB_2 (0,54-0,58, depending on the sintering temperature). For comparison, the coefficient of friction of the sintered AISI 316L steel was in the range of 0.67-0.72 (Figure 18a). The addition of TiB_2 improves the abrasion resistance of composites. This is the effect of the high hardness of TiB_2 ceramics, reaching 3400 HV [56]. TiB_2 particles protect the austenitic steel matrix during the process of friction, reducing the rate of wear. Therefore, in composites with lower content of TiB_2 (2 vol% and 4 vol%), the Al_2O_3 ball could penetrate the material of the steel matrix more easily and remove it during operation. Additionally, it was observed that higher sintering temperature reduced the coefficient of friction. The lowest values of the coefficient of friction were obtained for the material sintered at 1300°C. Similar correlations were observed in the specific wear rate of the tested composites (Figure 19a). The specific wear

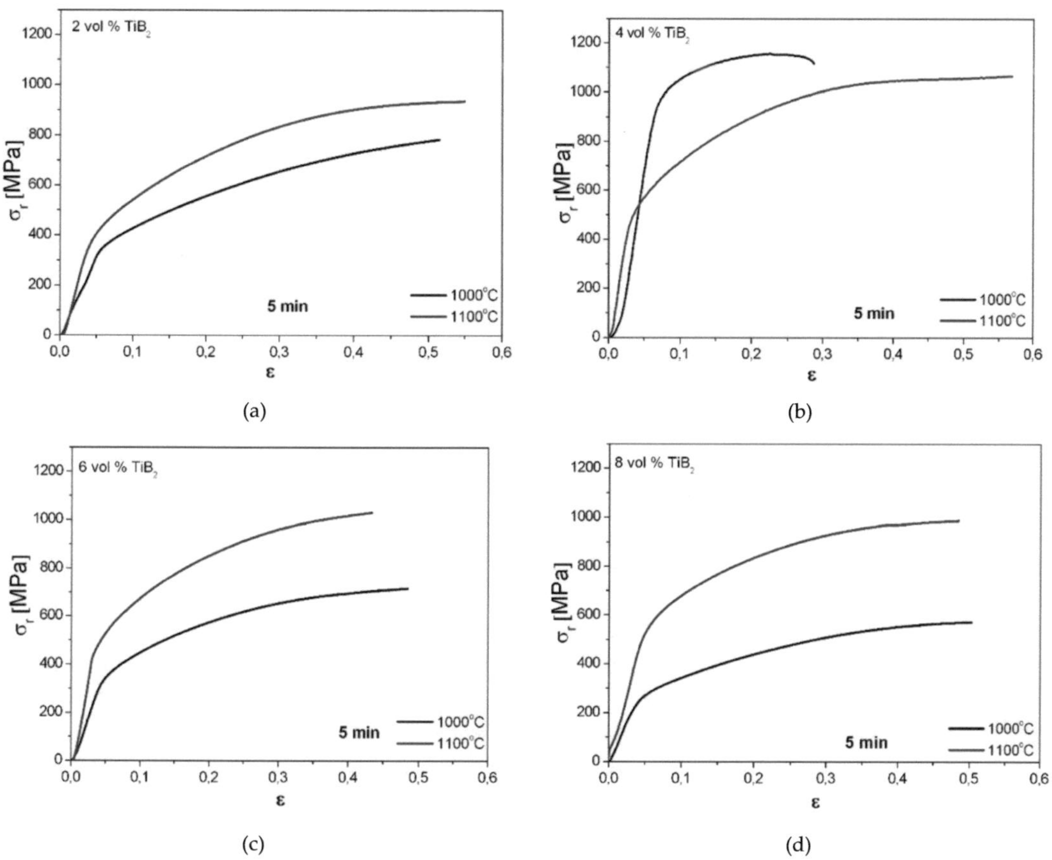

Figure 17. Compression strength of the composites with: a) 2 vol% of TiB_2 b) 4 vol% of TiB_2 c) 6 vol% of TiB_2 and d) 8 vol% of TiB_2 sintered by SPS method.

rate was decreasing with the increasing content of TiB_2 ceramics in the composite matrix. The specific wear rate also depended on the sintering conditions, which means that it was decreasing with the increasing temperature. The lowest value of the specific wear rate was obtained in a composite with 8 vol% of TiB_2 ($372*10^{-6}$ $mm^3/N \bullet m$), while in the AISI 316L steel, it amounted to $599*10^{-6}$ $mm^3/N \bullet m$ (Figure 19a).

Composites sintered by SPS (Figures 18b, 19b) offered improved abrasion resistance, compared with the materials sintered by HP-HT (Figures 18a, 19a). The lowest values of the coefficient of friction and of the specific wear rate were obtained in the materials manufactured by SPS at 1100°C, compared with the materials sintered by HP-HT at a temperature of 1300°C. In the case of these composites (SPS), adding the TiB_2 ceramics to the composite matrix considerably reduced the coefficient of friction (Figure 18b). The coefficient of friction assumed the highest value in the austenitic steels (0.62-0.61) and then was gradually decreasing to reach the lowest level in a composite with 8 vol% of TiB_2 (0.37). The obtained results prove that tribological properties depend on the parameters of the sintering process. Higher sintering temperature

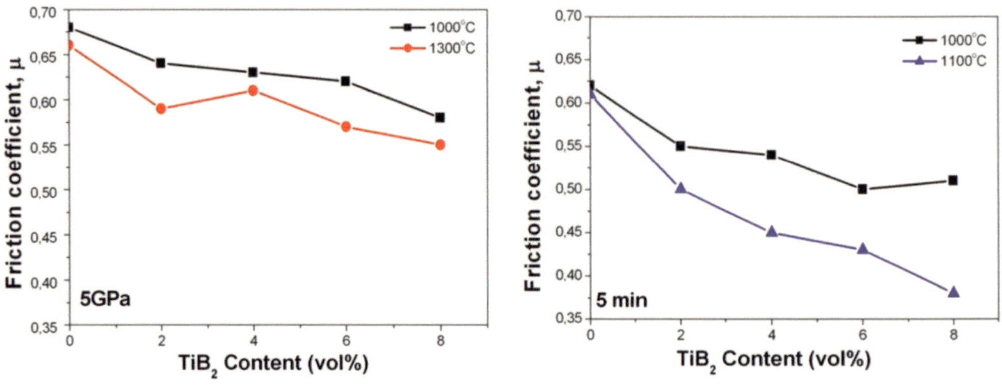

Figure 18. Variation of friction coefficient with TiB_2 content for composites sintered by: a) HP-HT [54] and b) SPS [53] method.

reduces the coefficient of friction, and thus the material wear rate. The best tribological properties were obtained in a material sintered at 1100°C for 5 minutes. For these parameters of sintering, the coefficient of friction assumed the values of 0.50, 0.45, 0.43 and 0.37 for composites containing 2 vol%, 4 vol%, 6 vol% and 8 vol% of TiB_2, respectively. By contrast, the specific wear rate was $429*10^{-6}$ [mm^3/Nm], $341*10^{-6}$ [mm^3/Nm], $314*10^{-6}$ [mm^3/Nm] and $299*10^{-6}$ [mm^3/Nm] for composites containing 2 vol%, 4 vol%, 6 vol% and 8 vol% of TiB_2, respectively. The results of tribological studies made on the composites fabricated by two different methods of sintering were consistent with the achievements of, among others, Tjong et al. [48,57]. The authors investigated the properties of composites reinforced with different volume fractions of the TiB_2 ceramics (5-20 vol%). Strong beneficial effect of the addition of TiB_2 particles on the abrasion resistance of austenitic stainless steel has been demonstrated.

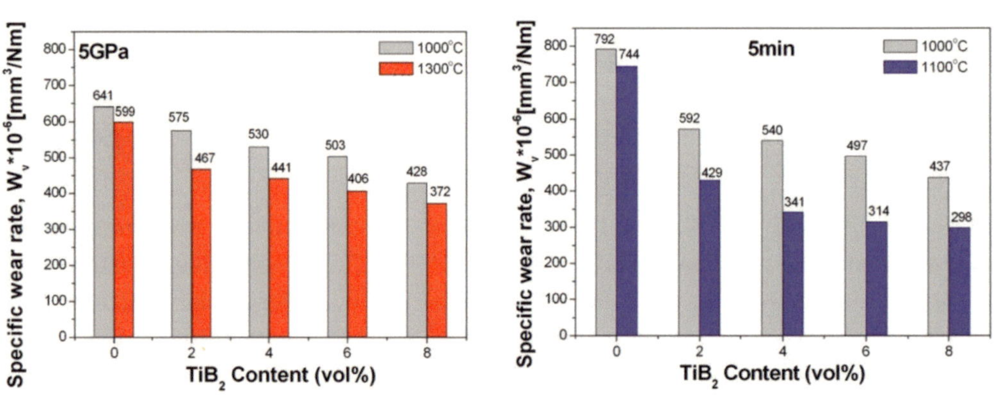

Figure 19. Variation of specific wear rate with TiB_2 content for composites obtained by: a) HP-HT [54] and b) SPS [53] method.

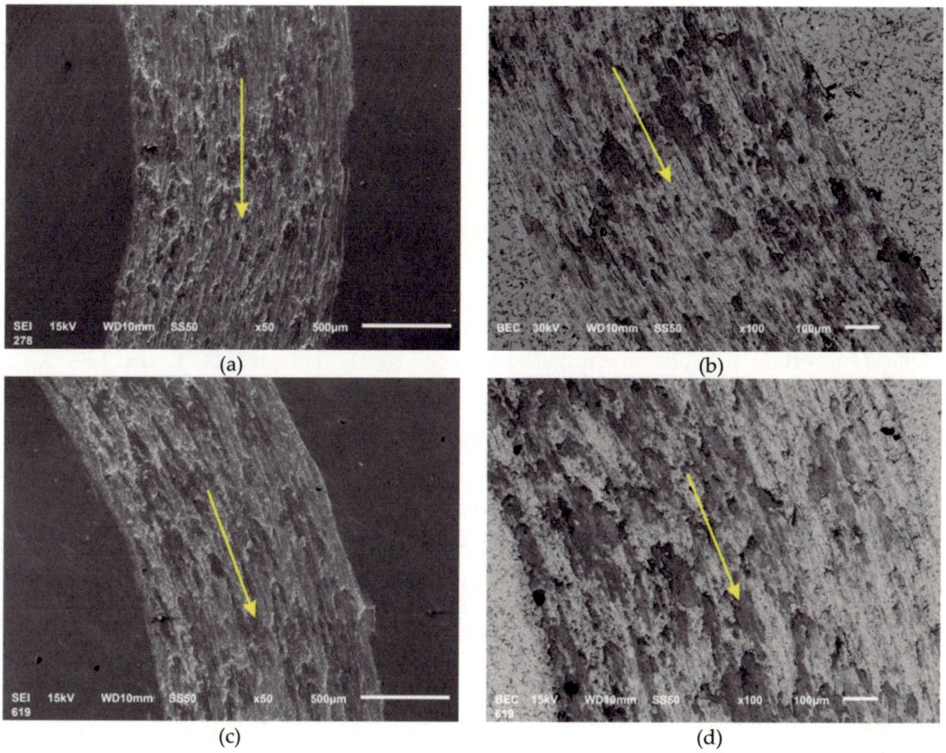

Figure 20. Selected SEM micrograph of the worn surface of composites with 8 vol% of TiB_2 obtained by: a, b) HP-HT (sintering temperature of 1300°C) and c, d) SPS (sintering temperature of 1100°C).

Microstructural examinations of wear traces formed in the process of abrasion followed the tribological tests (Figure 20). In all the tested samples, a very similar nature of the material removal by an Al_2O_3 ball was observed. There were visible scratches running parallel to the direction of the ball movement. Comparing wear traces in the composites studied, some characteristic features of the abrasive wear, such as scratches and grooves running in the direction of the ball movement, were observed (light-colour arrows in Figure 20). During abrasion process, permanent deformation of material combined with its attrition has occurred in the place of wear, but no cracks appeared on the worn out surfaces. Microstructural examinations (Figure 20b, d) of the places worn out by abrasion indicated the plastic nature of deformation. In all the tested composites, the surface of the worn out places showed the exposed particles of ceramics and local voids, probably caused by tearing the ceramic phase out of the matrix. The particles of TiB_2 present on the composite surface protected the matrix during abrasion and reduced its wear. This effect was particularly strong in composites reinforced with 6 vol% and 8 vol% of TiB_2.

4. Conclusions

Composite materials based on austenitic stainless steel with varying concentration of TiB_2 ceramics were fabricated. Four variants of the composites were obtained applying the technique of powder metallurgy and the two modern sintering processes, i.e. High Pressure-High Temperature (HP-HT) process and Spark Plasma Sintering (SPS). Based on the results obtained, it has been proved that the introduction of TiB_2 ceramics to the austenitic steel matrix is an efficient way to improve the composite properties. With increasing volume fraction of the reinforcing phase, an improvement in the physical, mechanical and tribological properties was observed. The best composite properties were obtained for the TiB_2 content of 8 vol%.

The study also showed a significant effect of the applied method of sintering on the properties of the tested materials. In all the composites sintered by HP-HT (at 1000°C and 1300°C, 5 GPa) and by SPS (only at 1100°C, 5 min), a very high degree of consolidation was obtained. It was the result of the combined effect of heat and high pressure in the case of the HP-HT method and of the use of electric discharges, high-rate heating and properly selected temperature in the case of the SPS process. Comparing the effect of the sintering methods used it can be concluded that the HP-HT process is a promising method for sintering the TiB_2-reinforced composites. Generally, the materials sintered at temperature of 1300°C and pressure of 5GPa were characterised by optimal microhardness and very good mechanical, plastic and tribological properties. In contrast, the use of SPS reduced the sintering temperature of composites, because already at a temperature of 1100°C and the duration of 5 min sinters with satisfactory density were produced. The best combination of physico-mechanical and tribological properties was obtained in the composites sintered at 1100°C. Careful analysis of the results has showed that the temperature of 1000°C was too low to produce a composite material with high degree of consolidation and satisfactory properties.

The sintering process carried out by either HP-HT or SPS has yielded the composite materials of a uniform and consistent structure produced within the whole volume of the sintered product. A homogeneous distribution of the reinforcing TiB_2 phase was obtained in the matrix of all the sintered products. The reinforcing phase showed a tendency to settle along the matrix grain boundaries. It has been demonstrated that the applied method of sintering and sintering conditions have a significant impact on changes in the microstructure of the tested composites. Application of the SPS process promotes the formation of new phases at the matrix boundary.

Acknowledgements

The author would like to thank Prof. Lucyna Jaworska and Paweł Figiel, Ph.D. and Piotr Putyra, Ph.D. from Institute of Advanced Manufacturing Technology in Cracow for help in SPS process of composites.

Author details

Iwona Sulima*

Address all correspondence to: isulima@up.krakow.pl

Institute of Technology, Pedagogical University of Cracow, Krakow, Poland

References

[1] Lis J., Pampuch R., Sintering, Uczelniane Wydawnictwa Naukowo-Dydaktyczne AGH, Kraków, 2000 (in polish).

[2] Kazior J., Analiza czynników technologicznych decydujących o własnościach spiekanych austenitycznych stali nierdzewnych, Monografia 164, Krakow, Wydawnictwo PK (1994) (in polish)

[3] Stuijts A. L., Synthesis of Materials from Powders by Sintering, Annual Review of Materials Science 1973; 3, 363-395.

[4] German R., Sintering: from Empirical Observations to Scientific Principles, Elsevier Inc. March, 2014.

[5] Kazior J.,' Nykiel M., Pieczonka T., Marcu Puscas T., Molinari A., Activated sintering of P/M duplex stainless steel powders, Journal of Materials Processing Technology, 2004; 157–158, 712–717.

[6] Bagliuk G, Properties and Structure of Sintered Boron Containing Carbon Steels, Sintering – Methods and Products, Edited by Dr. Volodymyr Shatokha, Rijeka, Intech; 2012, 249-266 (ISBN 978-953-51-0371-4).

[7] Celebi Efe G., Yener T., Altinsoy I., Ipek M., Zeytin S., Bindal C., The effect of sintering temperature on some properties of Cu–SiC composite, Journal of Alloys and Compounds, 2011; 509 (20), 6036–604.

[8] Duan X., Jia D., Wu Z., Tian Z., Yang Z., Wang S., Zhou Y., Effect of sintering pressure on the texture of hot-press sintered hexagonal boron nitride composite ceramics, Scripta Materialia, 2013; 68 (2) 104–107.

[9] Bundy F.P.; Ultra-high pressure apparatus, Physic Reports, 1988; 3, 156-175.

[10] Jaworska L., Wysokociśnieniowe spiekanie proszków diamentowych, Prace IOS, seria Zeszyty Naukowe Kraków, 82, 2002 (in polish).

[11] Eremts M.I., High pressure experimental method, Oxford University Press, 1996.

[12] Klimczyk P., Figiel P., Petrusha I., Olszyna A., Cubic boron nitride based composites for cutting applications, Journal of Achievements in Materials and Manufacturing Engineering 2011; 44 (2) 198-204.

[13] Sulima I., Figiel P., Suśniak M., Świątek M., Sintering of TiB_2 ceramic, Archives of Materials Science and Engineering, 2007; 28 (11) 687-690.

[14] Rozmus M. „Cermetalowe materiały gradientowe", Materiały Ceramiczne, 2006; 4,142-147 (in polish).

[15] Rozmus M., Jaworska L., Królicka B., Putyra P., Gradientowa mikrostruktura kompozytowych spieków diamentowych przeznaczonych na narzędzia skrawające, Materiały Ceramiczne /Ceramic Materials, 2009; 61 (3) 192-196 (in polish).

[16] Wyżga P., Jaworska L., Bućko M., Putyra P., Kalinka A., Sintering of TiB_2-TiN nano- and micropowders, Composite Theory and Practice, 2011; 1, 34-38.

[17] Sulima I., Klimczyk P., Hyjek P., The influence of the temperature and pressure on the properties of the AISI 304L stainless steel reinforced with TiB_2 ceramic, Archives of Materials Science and Engineering, 2009; 39 (2)103-106.

[18] Sulima I., Jaworska L., Wyżga P., Perek-Nowak M.,The influence of reinforcing particles on mechanical and tribological properties and microstructure of the steel-TiB_2 composites, Journal of Achievements in Materials and Manufacturing Engineering, 2011; 48 (1) 52-57.

[19] Figiel P., Jaworska L., Putyra P., Klimczyk P., Bryła K., Wysokociśnieniowe i swobodne spiekania kompozytów cermetalowych z udziałem nanometrycznych proszków TiC, Composite Theory and Practice, 2008, 130-135 (in polish).

[20] Taylor G.F.: Apparatus for Making Hards Metal Compositions, U.S. Patent No. 1896854, 1933

[21] Orru R., Licheri R., Cincotti A. M., Cao G.: Consolidation/synthesis of materials by electric current activated/assisted sintering, Materials Science and Engineering 2009; 63, 127-134.

[22] Inoue K., US Patent No. 3 24,1956

[23] Omori M., Sakai H., Okubo A., Kawahara M., Tokita M., Hirai T., Preparation and Properties of ZrO_2(3Y)/Ni FGM, Proceedings of the 3rd International Symposium on Structural and Functionally Gradient Materials, Lausanne, Switzerland, 1994; 99-104.

[24] Omori M., Sakai H., Okubo A., Tokita M., Kawahara M., T. Hirai, Preparation of Functional Gradient Materials by Spark Plasma Sintering, Symposium of Materials Research Society of Japan, 1994.

[25] Yoshimura M., Ohji T., Sando M. Nihara K., Rapid rate sintering of nano-grained ZrO_2-based composites using pulse electric current sintering method. Journal of Materials Science Letters, 1998; 17, 1389-1391.

[26] Tokita M.: Mechanism of Spark Plasma Sintering. 1996 http://xa.yimg.com/kq/groups/3862917/2054596553/name/SUMITOMO+REVIEW-Spark-Plasma-Sintering.pdf.

[27] Omori M., Sintering, consolidation, reaction and crystal growth by the spark plasma system (SPS), Materials Science Engineering, 2000; A287; 183-188.

[28] Peng H., Spark Plasma Sintering of Si_3N_4-Based Ceramics-Sintering mechanism-Tailoring microstructure-Evaluating properties, PhD thesis, Department of Inorganic Chemistry, Stockholm University, 2004.

[29] Tokita M., Trends in Advanced SPS Spark Plasma Sintering Systems and Technology, Journal of the Society of Powder Technology Japan, 1993; 30 (11) 790-804.

[30] Tokita M., Mechanism of spark plasma sintering, Proceeding of NEDO International Symposium on Functionally Graded Materials, Kyoto, Japan, 1999, 23-33.

[31] Zang J., Field Activated Sintering Technology: Multi-physics Phenomena Modelling, A Coupled Thermal-electrical-densification Framework, Lambert Academic Publishing, Saarbrücken, 2010.

[32] Putyra, P., Figiel P., Podsiadło M., Klimczyk P., Alumina composites with solid lubricant participations, sintered by SPS-method, Kompozyty, 2011; 2, 107-110.

[33] Pellizzari M., Fedrizzi A., Zadra M., Influence of processing parameters and particle size on the properties of hot work and high speed tool steels by Spark Plasma Sintering, Materials and Design, 2011; 32, 1796–1805.

[34] Venkateswaran T., Basu B., Raju G.B., Kim D.Y., Densification and properties of transition metal borides-based cermets via spark plasma sintering, Journal of the European Ceramic Society 2006; 26, 2431–2440.

[35] Dustin D. M., Dongtao J., Dina V. D., Mukherjee A. K., The synthesis and consolidation of hard materials by spark plasma sintering, International. Journal of Refractory Metals & Hard Materials, 2009; 27, 367–375.

[36] Material Science and Engineering Handbook, Ed. James F. Shackelford, Second Edition by CRC Press, Florida, 2001.

[37] Raju G.B, Mukhopadhyay A., Biswas K., Basu B., Densification and high-temperature mechanical properties of hot pressed TiB_2–(0–10 wt.%) $MoSi_2$ composites, Scripta Mater. 2009; 61, 674-6.

[38] Konigshofer R., Furnsinn S., Steinkellner P., Lengauer W., Haas R., Rabitsch K., Scheerer M., Solid-state properties of hot-pressed TiB_2 ceramics, International Journal of Refractory Metal and Hard Materials, 2005; 23 350–357.

[39] Wang F. C., Zhang Z. H., Luo J., Huang C. C., Lee S. K., A novel rapid route for in situ synthesizing TiB-TiB_2 composites, Composites Science and Technology, 2009; 69 (15-16) 2682–2687.

[40] Matkovich, V. I., Boron and Refractory Borides. Springer, Berlin, 1977; 172.

[41] Balcı Ö., Ağaoğulları D., Gökçe H., Duman İ., Öveçoğlu M. L, Influence of TiB_2 particle size on the microstructure and properties of Al matrix composites prepared via mechanical alloying and pressureless sintering, Journal of Alloys and Compounds, 2014; 586 (1) S78–S84.

[42] Kwon Y.-S., Kim H.-T., Kim j.-S., Dudina D.V., Spark plasma sintering of Cu-TiB_2 nanocomposite, Novel Materials Processing by Advanced Electromagnetic Energy Sources, Proceedings of the International Symposium on Novel Materials Processing by Advanced Electromagnetic Energy Sources March 19–22, Osaka, Japan, 2005; 293–296.

[43] Jaroszewicz J., Michalski A., Preparation of a TiB_2 composite with a nickel matrix by pulse plasma sintering with combustion synthesis, Journal of the European Ceramic Society, 2006; 26 (13) 2427–2430.

[44] Animesh A., Bandyopadhyay T.K., Das K., Synthesis and characterization of TiB_2-reinforced iron-based composites, Journal of Materials Processing Technology, 2006; 172, 70–76.

[45] Tjong S.C., Tam K., F., Mechanical and thermal expansion behaviour of hipped aluminium-TiB_2 composites", Materials Chemistry and Physics, 2006; 97, 91-97.

[46] Pettersson A., Magnusson P., Lundberg P., Nygren M., Titanium–titanium diboride composites as part of a gradient amour material, International Journal of Impact Engineering, 2005; 32, 387–399.

[47] Nahme H., Lach E., Tarran A., Mechanical property under high dynamic loading and microstructure evaluation of a TiB2 particle-reinforced stainless steel, Journal of Materials Science, 2009; 44, 463-468.

[48] Tjong S.C., Lau K.C. Abrasion resistance of stainless-steel composites reinforced with hard TiB_2 particles, Composites Science and Technology, 2000; 60 (8) 1141-1146.

[49] International Standard, Fine ceramics (advanced ceramics, advanced technical ceramics)-Determination of friction and wear characteristics of monolithic ceramics by ball-on-disc method, ISO 20808:2004(E).

[50] Meozzi M., Special use of the ball on disc standard test, Tribology International, 2006; 39 (6) 496–505.

[51] Sulima I., Figiel P., Kurtyka P., Austenitic stainless steel –TiB_2 composites obtained by HP-HT method, Composites Theory and Practice, 2012; 12 (4) 245-250.

[52] Sulima I., L.Jaworska, P.Figiel, Influence of processing parameters and different content of TiB_2 ceramics on the properties of composites sintered by high temperature – high pressure (HT-HP) method, Archives of Metallurgy and Materials, 2014; 59 (1) 203-207.

[53] Sulima I. Tribological properties of steel/TiB_2 composites prepared by spark plasma sintering, Archives of Metallurgy and Materials (2014; in press).

[54] Sulima I, P. Klimczyk, P.Malczewski, Effect of TiB_2 particles on the tribological properties of stainless steel matrix composites, Acta Metallurgica Sinica (English Letter), 2014; 27(1), 12–18.

[55] CRC Materials Science and Engineering Handbook, Third Edition Edited by James F. Shackelford and William Alexander CRC Press, 2001, 509.

[56] McGuire M. F., Stainless Steels for Design Engineers, ASM International, 2008, 69-78.

[57] Tjong K.C., Lau K.C., Sliding wear of stainless steel matrix composite reinforced with TiB_2 particles, Materials Letters, 1999; 4 (4) 153-158.

Chapter 8

Correlation between Thermal and Electrical Properties of Spark Plasma Sintered (SPS) Porous Copper

Yaniv Gelbstein, Yedidia Haim, Sergei Kalabukhov, Vladimir Kasiyan, S. Hartmann, S. Rothe and Nahum Frage

Additional information is available at the end of the chapter

http://dx.doi.org/10.5772/59010

1. Introduction

Spark Plasma Sintering (SPS) is a consolidation approach, in which pulsed direct current passes through a graphite die and a compacted specimen, and enabling very high heating rates with simultaneously applying an external pressure. The combination of internal heating and external pressure provides the conditions for fast sintering. SPS parameters have to be determined based on the thermal and electrical properties of the consolidated material, which, evidently, depend on the compact porosity that decreases during the sintering process. In the present study, the effect of porosity of the SPS-processed Cu specimens on their thermal and electrical properties at room temperature was theoretically and experimentally investigated.

There are some theoretical approaches, which may be applied for analysis of transport properties of a porous material. One of the approaches, the "so-called" general effective media (GEM) method (eq. 1) was discussed in [1, 2].

$$x_1 \cdot \frac{(\sigma_1,\kappa_1)^{1/t} - (\sigma_{eff},\kappa_{eff})^{1/t}}{(\sigma_1,\kappa_1)^{1/t} + A\cdot(\sigma_{eff},\kappa_{eff})^{1/t}} = (1-x_1)\frac{(\sigma_{eff},\kappa_{eff})^{1/t} - (\sigma_2,\kappa_2)^{1/t}}{(\sigma_2,\kappa_2)^{1/t} + A\cdot(\sigma_{eff},\kappa_{eff})^{1/t}} \quad (1)$$

The GEM equation (eq. 1) is usually employed for calculating the effective electrical and thermal conductivities (σ_{eff} and κ_{eff}, respectively) for two-phase materials using the electrical (σ_1 and σ_2) and thermal (κ_1 and κ_2) properties of each phase. The morphological (or geometrical) parameters (A and t) may be derived from the equation of conductivity percolation [1] or by appropriate modeling of experimental results. The x_1 and (1-x_1) values are the volume fractures of the phases in a two-phase material.

Generally, A and t values depend on the phases distribution and their morphology within two-phase materials. Let us consider a composite material, which consists of continuous matrix and homogenous distributed particles with high aspect ratio (for instance, fibers) of a second phase. For this kind of distribution t value is equal to 1 [1] and the so-called "parallel" and "series" alignment of the particles (relative to the electrical potential or temperature gradients) may be considered. The parameter A varies from ∞ for the parallel to 0 for series alignments. Recently [1], the GEM equation and measured transport properties were successfully used for estimating the fracture and distribution of Sn phase, which displays a fiber shape particles, in the Sn_xTe_{1-x} inter-metallic matrix.

Ke-Feng Cai et al. [3] reported the results of prediction of effective thermal conductivity of porous Al-doped SiC ceramics using the Landauer's expression (eq. 2) derived for spherical and homogeneously dispersed second phase [4].

$$\kappa_{eff} = \kappa_1 \left(\frac{3x_1 - 1}{2} \right) \tag{2}$$

It may be shown that this equation may be derived from the GEM equation (1) if one takes into account that A=2 and t=1.

Another commonly used approach, yet not straightforward for multi-phased materials, to correlate effective transport properties (thermal and electrical conductivity) of porous materials is based on the Weidemann-Franz relation.

$$\kappa_{eff} = L \cdot \sigma_{eff} \cdot T \tag{3}$$

Where, the coefficient L is known as Lorenz number and T represents the absolute temperature. The Lorenz number is constant only in the case when the conduction electrons are scattered elastically. This condition requires that the temperature-independent electron scattering by impurities will dominate over the electron scattering by phonons [5]. These conditions may be achieved at high temperatures or at very low temperatures, where the residual resistivity is predominant. At intermediate temperatures, the condition of elasticity no longer holds and the Lorenz number decreases considerably from the Sommerfeld value. In order to get a better understanding of the L number, let us consider the general expression for L, derived from the Fermi-Dirac statistics (eq. 4) [6] and graphically presented in Fig. 1 for two different electron scattering mechanisms (or scattering parameters, r): by ionized impurities (using scattering parameter r of 3/2) and by acoustic phonons (r=-1/2).

$$L = \left(\frac{k}{e}\right)^2 \left[\frac{\left(r+\frac{7}{2}\right)\left(r+\frac{3}{2}\right) F_{r+5/2}(\eta) F_{r+1/2}(\eta) - \left(r+\frac{5}{2}\right)^2 F^2_{r+3/2}(\eta)}{\left(r+\frac{3}{2}\right)^2 F^2_{r+1/2}(\eta)} \right] \tag{4}$$

Where, e, k, F_r and η are the electrons charge, Boltzmann constant, Fermi integral (defined by eq. 5) and the reduced Fermi energy (E_F/kT), respectively.

$$F_r = \int_0^\infty \xi^r f_o(\eta) \cdot \partial\xi \tag{5}$$

$$f_o(\eta) = \frac{1}{1+\exp(\xi - \eta)} \tag{6}$$

In the expression for F_r (eq. 5)-f_o (defined in eq. 6) and ξ are the Fermi distribution function and the kinetic energy of a charge carrier, respectively.

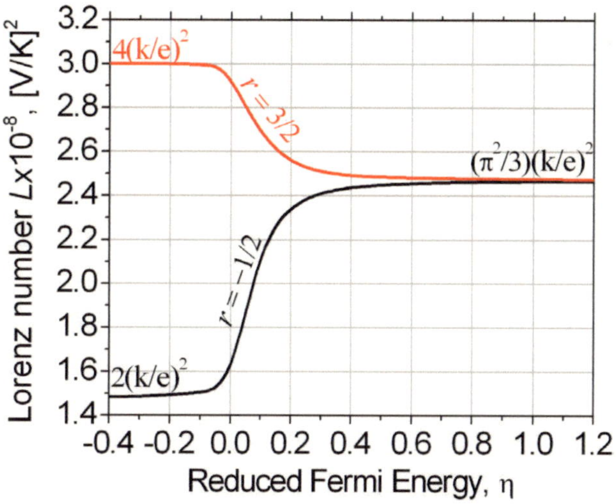

Figure 1. Lorenz number variation with the reduced Fermi energy (η=E_F/kT) for two different electron scattering mechanisms, by ionized impurities (using scattering parameter r of 3/2) and by acoustic phonons (r=-1/2).

According to this analysis, for materials with low carrier concentrations (η<-0.4), L values reach the classical (Boltzmann statistics) values of $4(k/e)^2$ and $2(k/e)^2$ for ionized impurities and acoustic phonons scattering, respectively. For degenerate materials (η>1.2) L reaches the Sommerfeld value, derived from eq. 7, regardless the scattering mechanism.

$$L = \frac{\pi^2}{3}\left(\frac{k}{e}\right)^2 \tag{7}$$

For the intermediate η values, *L* exhibits a strong dependence on both Fermi energy and the scattering mechanism.

Since Cu is highly conductive metal, the value of η at room temperature may be calculated using eq. 8 and the physical properties of Cu (effective mass, m^*=1.01m_o [7], and carrier concentration, n=8.4x10^{22} cm^{-3} [8]). The estimated η value is equal to 270 and corresponds to the energetic range where the constant Sommerfeld value valids.

$$n = \frac{8}{3\sqrt{\pi}}\left(\frac{2\pi m^* kT}{h^2}\right)^{3/2} \eta^{3/2} \tag{8}$$

Nevertheless, the Lorenz number differs slightly from the Sommerfeld value due to the inelastic electrons scattering.

The simple consideration presented above shows that for the applying the Weidemann-Franz relation between the transport properties of materials, the specific Lorenz number has to be determined.

It was shown [9], for instance, that Lorenz number of copper alloys depends on their purity and thermo-mechanical treatments. Moreover, in [10] it was established that the Lorenz number for copper films depends on its thickness. This effect was partly attributed to scattering of electrons at the films surfaces and partly to scattering by frozen-in structural defects.

This explanation is based on a distinction between the free paths of electrons for the electrical conductivity process and those for the thermal conductivity process. Being more specific, in a metal, an electric field or a temperature gradient causes an electron drift, which is restricted only by the collisions of the electrons with lattice imperfections (static defects or lattice vibrations). When the electron distribution function is disturbed from its equilibrium value, the rate of return to equilibrium may be expressed by collision processes, which are usually expressed in terms of a relaxation time. Only in case that the relaxation time is the same for both electrical and thermal transport, eq. 7 can be used. This equation is based on the assumption that *L* is a constant independent of the band structure or the relaxation time. Regarding relaxation, it was pointed out in [11] that equilibrium can be reached in two ways: either by processes changing the direction of motion of an electron but not changing its energy significantly, or by processes changing its energy but not the direction – the so-called "horizontal" or "vertical" movements on the Fermi surface. Since the "vertical" movement was found as ineffective in producing electrical resistance, the relaxation times for electrical and thermal conduction are equal only in case that the "vertical" movement is absent. The effective scattering by phonons at high temperatures and by impurities at low temperatures is elastic, leading to similar relaxation time for the different transport properties. This is the reason for the validity of eq. 7 at very low or high temperatures.

Koh et al. [12] had used a modified Weidemann-Franz relation (eq. 9), which takes into account also the lattice component of the thermal conductivity, for analysis of porous stainless steel and Cu based alloys.

$$\kappa_0 = L\frac{T}{\rho_0} + b \tag{9}$$

Where, ϱ_0 and κ_o are the electrical resistivity and the thermal conductivity of the alloys, respectively, and b is the lattice component of the thermal conductivity.

It was established that for highly conductive materials, such as copper alloys, where the lattice component of the thermal conductivity is relatively small compared to the electronic one, the dependency of the thermal conductivity on T/ϱ for various porosity levels was characterized by a straight curve with a slope L and intercept b. In this case the L and b values depend on the nature of metal only. For stainless steel, for which the lattice and the electronic components are comparable, b value depends on the porosity and specific curve for each porosity level was obtained.

It can be concluded that special care should be taken while using the Weidemann-Franz relation for porous or other multi-phased materials. On the other hand, applying the GEM approach is much more straightforward, giving additional information about the phases' alignment and distribution characteristics

In the presence study, the GEM and Weidemann-Franz relations were applied for investigation of the experimental transport properties results of porous SPS-processed Cu specimens.

2. Experimental

Pure (99.9%) copper powder with a nearly spherical particles shape and an average particle size of about 8μm was consolidated by SPS apparatus (FCT Systems GmbH, Germany) under argon atmosphere with heating and cooling rates of 50°C/min. The specific parameters of the SPS process are presented in Table 1. The porosity of the sintered specimens, (measured by the Archimedes method) was varied in the 0-30 vol.% range.

The microstructure was characterized by optical microscope *Zeiss* (Germany). Electrical resistivity was measured at room temperature by a four-probe method using 1V/50 Hz alternating power source and *Keithley 2182A Nanovoltmeter*. The thermal conductivity was tested at room temperature using the flash diffusivity method (*LFA 457, Netzsch*). Thermal conductivity (κ_{eff}) values were calculated using the equation $\kappa_{eff}=\alpha\, \varrho\, C_p$ where, α is the thermal diffusivity, C_p is the specific heat (measured using differential scanning calorimetry, *STA 449-Netzsch*), and ϱ is the bulk density of the sample.

Porosity, vol%	Temperature, °C	Holding time, min.	Applied uniaxial pressure, MPa
0	700	15	97
1	600	15	97
10	400	2	88
15	400	2	70
20	400	2	53
30	400	2	35

Table 1. Spark Plasma Sintering conditions and porosity of the specimens.

The thermal conductivity measurements were conducted under ~1atm argon, similarly to the SPS conditions.

3. Results

The microstructure of the porous Cu speciemns is shown in Fig. 2.

The average grain size of about 7-8μm indicates no coarsening effects during the SPS process. The pores are homogenously distributed within the metallic matrix.

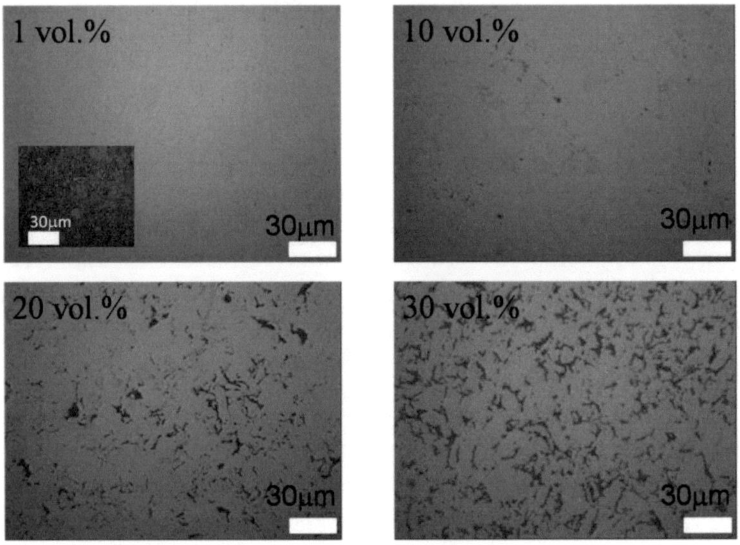

Figure 2. Optical images of the SPS-processed specimens. The microstructure of the almost dense specimen (1vol% porosity) after etching was presented as the insert in the corresponded image.

Values of the electrical resistivity and thermal conductivity measured at room temperature as a function of porosity are shown in Fig. 3.

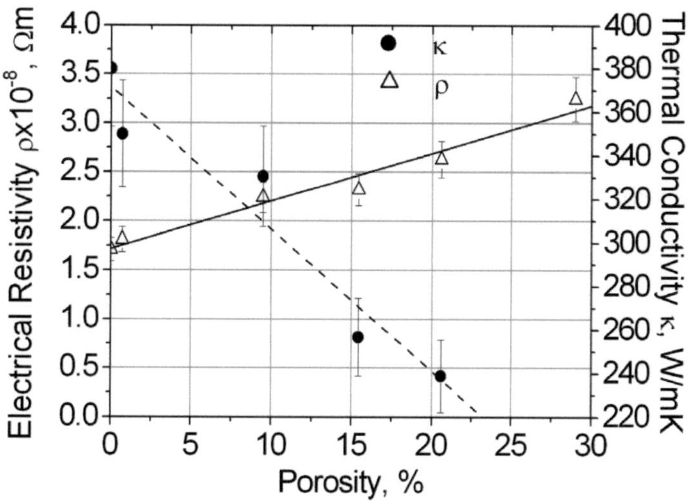

Figure 3. The electrical resistivity and thermal conductivity of porous Cu specimens.

As was expected, a general trend of increasing the electrical resistivity and decreasing the thermal conductivity values with increasing of the porosity amount was observed. Furthermore, additional insights regarding the geometrical alignment and porous distribution can be obtained, upon applying the GEM and the Weidemann-Franz equations.

4. Discussion

In order to apply the GEM equation (1) and Weidemann-Franz relation (eq. 3) for analyzing the experimental results presented in Fig. 3, the thermal and electrical conductivities of pure Cu and argon, which filled the micrometric pores, have to be determined. The thermal and electrical conductivities of pure copper (κ_1 and σ_1, respectively, in eq. 1) are reported in [13, 14]. For argon, the electrical conductivity, σ_2, is negligible, while the thermal properties has to be discussed. According to [15], at normal conditions, heat convection through a porous metallic structure becomes of practical importance only at temperatures above 1000K and for pore diameter larger than 5000μm. It was also reported in [15] that heat transfer by radiation in the pores at moderate and high temperatures might be neglected. Thus, for materials with relatively small pores, as in our case, conductive heat transfer is the dominant process at the investigated temperature.

Under normal pressure, the thermal conductivity of Ar [16, 17] may be calculated according to eq.10

$$\kappa_{Ar} = 1.47 \cdot 10^{-5} \frac{T^{1/2}}{1 + 142/T} \quad [Watt/cmK] \tag{10}$$

and is equal to ~1.73x10^{-2} W/mK at room temperature.

4.1. General Effective Media (GEM) theory

The aims of this paragraph is to confirm that the effective transport properties of the porous material may be calculated using one set of the geometrical parameters A and t. We started with the analysis of the measured values of the effective transport properties for the specimen showing a 20 vol.% porosity. By solving together two equations, based on the GEM approach, for the thermal and electrical conductivities, the values A=1.3 and t=1 were obtained. These values were used for calculation of the transport properties for the entire investigated specimens, showing various porosity amounts. A very good agreement was found between the calculated and experimental results, as can be seen in Figs. 4-5, while comparing the experimental data to various calculated curves with different A values. The experimental points are located between the curves corresponded to A values in the range of 1÷2. Thus, we may conclude that the pores are homogeneously dispersed and have nearly spherically shaped, in agreement with the micro-structural observations (Fig. 2).

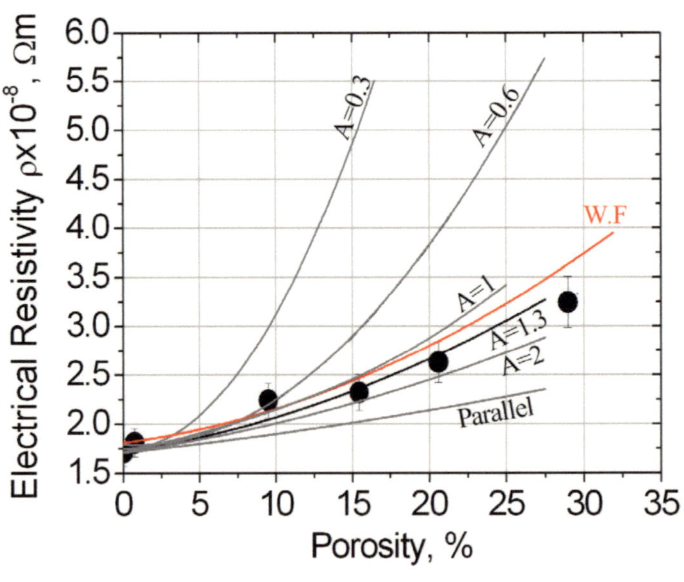

Figure 4. Effective electrical resistivity (=σ_{eff}^{-1}) values for the investigated porous Cu samples, calculated by the GEM equation substituting A and t parameters of 0.3,1; 0.6,1; 1.3,1; 2,1 and ∞,1 (parallel alignment); and by the Weidemann-Franz (W.F.) relation. The experimentally measured results are shown by the black points.

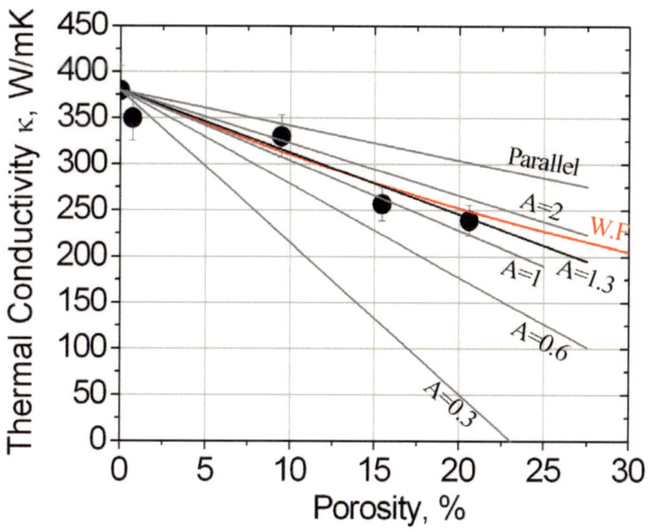

Figure 5. Effective thermal conductivity (κ_{eff}) values for the investigated porous Cu samples, calculated by the GEM equation substituting *A* and *t* parameters of 0.3,1; 0.6,1; 1.3,1; 2,1 and ∞,1 (parallel alignment); and by the Weidemann-Franz (W.F.) relation. The experimentally measured results are shown by the black points.

4.2. Weidemann-Franz (W.F.) relation

The calculated values of the transport properties as a function of the porosity levels using Weidemann-Franz relation (eq. 3) are also presented in Figs 4-5. The calculations took into account the Lorenz number for Cu at 0°C (2.23x10^{-8} [V/K]^2) reported in [18]. This value is slightly differs from the Sommerfeld value of 2.47x10^{-8} [V/K]^2 (derived from eq. 7), which is commonly used for highly degenerated materials. Nevertheless, there is a good agreement between the calculated and the experimental results for both the electrical resistivity and the thermal conductivity values.

4.3. W.F. vs. GEM

While comparing the two suggested approaches for estimating the electrical and thermal transport properties of porous materials, several general guidelines can be obtained:

1. Applying the GEM approach upon measurement of κ_{eff} and σ_{eff} (or ϱ_{eff}) of one known porosity level, and previous knowledge of the same properties of the matrix phase, can give a good approximation for the porosity amount dependency of the electrical and thermal conductivities for any given porosity amount in addition to some insight about the porosity alignment, distribution and morphology. It should be remembered that, a special care should be taken upon using any approximation such as eq. 2, since dependent on the powdering approach and the crystallographic nature of the matrix composition, the porosity morphology do not have to be spherical with an isotropic distribution.
2. Applying the Weidemann-Franz relation (equation 3) for a metallic or highly degenerated porous material, using the Sommerfeld value of *L* (derived from eq. 7) at moderate

temperatures (e.g. room temperature) leads to inaccurate estimations due to inelastic scattering effects. Therefore, measurement of both κ_{eff} and σ_{eff} (or ϱ_{eff}) for one known porosity level, for evaluation of L, should not be avoided. For less conductive matrix materials, phonons thermal conductivity effects are involved, which are expected to vary with the porosity amount. For semiconductors with reduced Fermi energy in the range of $-0.4<\eta<1.2$, L depends also on the scattering mechanism and the Fermi energy. For semiconductors or insulators with $\eta<-0.4$, L depends on the scattering mechanism as well. Such an approach can't supply any insight on the geometrical alignment and morphology of the pores.

5. Conclusions

The thermal and electrical conductivities for spark plasma sintered porous Cu samples were experimentally measured at room temperature and theoretically correlated to the GEM and Weidemann-Franz relations. Both of the theoretic approaches showed a good agreement to the experimental values. Yet several restrictions were discussed upon generalizing the proposed theoretical procedures for any porous material. Applying the GEM equation was found as the ideal approach for estimating the thermal and electrical characteristics of porous materials, due to the possibility to give a good approximation for the porosity amount dependency of the electrical and thermal conductivities for any given porosity amount in addition to some insight about the porosity alignment, distribution and morphology.

Acknowledgements

The authors wish to acknowledge Mr. Yair George for his assistance in the experimental measurements.

Author details

Yaniv Gelbstein[1*], Yedidia Haim[2], Sergei Kalabukhov[1], Vladimir Kasiyan[1], S. Hartmann[3], S. Rothe[3] and Nahum Frage[1]

*Address all correspondence to: yanivge@bgu.ac.il.

1 Department of Materials Engineering, Ben-Gurion University of the Negev, Israel

2 Department of Mechanical Engineering, Ben-Gurion University of the Negev, Israel

3 Institute of Applied Mechanics, Clausthal University of Technology, Clausthal-Zellerfeld, Germany

References

[1] Yaniv Gelbstein, "Thermoelectric Power and Structural Properties in Two Phase Sn/SnTe Alloys", Journal of Applied Physics 105, 023713 (2009).

[2] Maclachlan, D.S., Blaszkiewicz, M. &Newnham, R.E. *J Am Ceram Soc*73(8), 2187-2203 (1990).

[3] Ke-Feng Cai, Jin-Ping Liu, Ce-Wen Nan, Xin-Min Min, "Effect of porosity on the thermal-electric properties of Al-doped SiC ceramics", Journal of Materials Science Letters 16 (1997) 1876-1878.

[4] Rolf Landauer, "The Electrical Resistance of Binary Metallic Mixtures", Journal of Applied Physics Vol. 23, No. 7, p.779 (1952).

[5] R. W. Arenz, C. F. Clark, and W. N. Lawless, "Thermal conductivity and electrical resistivity of copper in intensemagnetic fields at low temperatures", Physical Review B Vol. 26, No. 6, p. 2727 (1982).

[6] Y. Gelbstein, Z. Dashevsky and M.P. Dariel, "High performance *n*-type PbTe-basedmaterials for thermoelectric applications", Physica B 363 196-205 (2005).

[7] Arthur Beiser, "Perspectives of Modern Physics", Mc-Graw-Hill International Editors, p.487 (1969).

[8] Edward Ramsden, "Hall-Effect Sensors – Theory and Applications", 2nd Edition, Elsevier, p. 8 (2006).

[9] Robert L. Powell, Hans M. Roder, and William J. Hall, "Low-Temperature Transport Properties of Copper and Its Dilute Alloys: Pure Copper, Annealed and Cold-Drawn", Physical Review Vol. 115, No. 2, p. 314 (1959).

[10] Prem Nath and K. L. Chopra, "Thermal Conductivity of Copper Films", *Thin Solid Films, 20* (1974) 53-62.

[11] G. K. White and R. J. Tainsh, "Lorenz Number for High-Purity Copper", Physical Review Vol. 119, No. 6, p.1869 (1960).

[12] J. C. Y. Koh and Anthony Fortini, "Prediction of Thermal Conductivity and Electrical Resistivity of Porous Metallic Materials", Int. J. Heat Mass Transfer Vol. 16, pp. 2013 (1973).

[13] R.P. Tye, "Thermal Conductivity", vol. 2, p.147, Academic Press (1969).

[14] Donald R. Askeland, "The Science and Engineering of Materials", 3rd edition, p. 597, PWS Publishing Company (1989).

[15] R. Zabbarov, "Effect of Porosity and Treatment on the Thermal Conductivity of Cermets", Journal of Engineering Physics 213 (1967), *(translated from: Inzhenerno-Fiziehe-skii Zhurnal, Vol. 13, No. 3, pp. 373-375, 1967).*

[16] Saul Dushman, "Scientific Foundations of Vacuum Technique", 2nd edition, pp. 40, 42, John Wiley & Sons inc. (1962).

[17] D.W. Stops, "The Mean Free Path of Gas Molecules in the Transition Regime", J. Phys. D: Appl. Phys. 3 685 (1970).

[18] N.F. Mott and H. Jones, "The Theory of the Properties of Metals and Alloys", Oxford University Press, p.307 (1958).

Section 4

Ram Extrusion and Hot Sintering of Reprocessed PTFE

Chapter 9

Recycling of Polytetrafluoroethylene (PTFE) Scrap Materials

Arunachalam Lakshmanan and S.K. Chakraborty

Additional information is available at the end of the chapter

http://dx.doi.org/10.5772/59599

1. Introduction

1.1. Plastics

Polymers are formed from *thermoplastic* and *thermosetting plastic* materials. The binding forces between polymer chains in thermoplastics such as polyethylene are the result of van der Walls forces between the molecules and mechanical entanglement between the chains as shown in Fig.1. Most of the thermoplastics can be reused after melting since the bonds between the molecules are easily broken on heating. However, in thermosetting plastics such as Bakelite various polymer chains are held together by strong covalent bond. They are rigid, strong and more brittle. Due to strong covalent bond and cross-link, they are insoluble in almost all organic solvents. They will not become plastic when heated.

1.2. PTFE

Polytetrafluoroethylene (PTFE) was discovered by a research chemist in DuPont in the year 1938. In 1941 it has been patented and got the first brand name as Teflon. It is a fluorinated polymer obtained from tetrafluroethylene (TFE) monomer through free radical vinyl polymerization. Tetra means four carbon atoms are covalently bonded to carbon atoms. Fluro means bonded atoms are fluorine. Ethylene means carbon atoms are joined by a double bond as in the case of ethylene.

In PTFE, carbon to carbon atom double bond becomes a single bond and a linear chain of carbon atoms are formed with two fluorine atoms covalently bonded to each carbon atom. These fluorine atoms shield the carbon atoms and hence no solvent can attack the carbon atoms. As a result, PTFE exhibits extraordinary chemical resistance to acids and alkalis. Carbon to fluorine bonds have high dissociation energy. Due to the high electronegativity of fluorine,

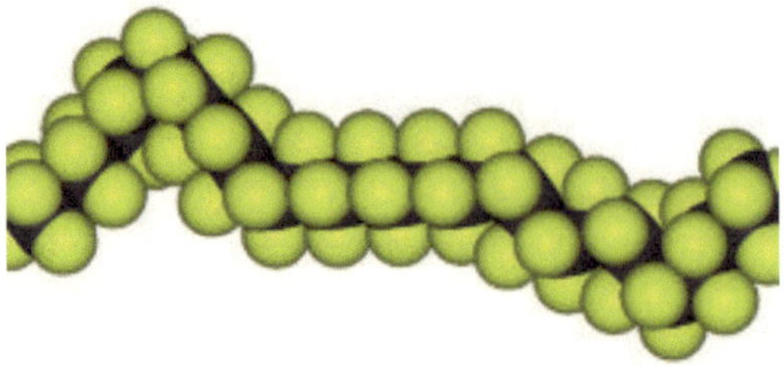

Figure 1. Typical interwinded PTFE polymer. Due to strong intermolecular forces, the polymer chains are tangled. Due to its chemical inertness, PTFE cannot be cross-linked.

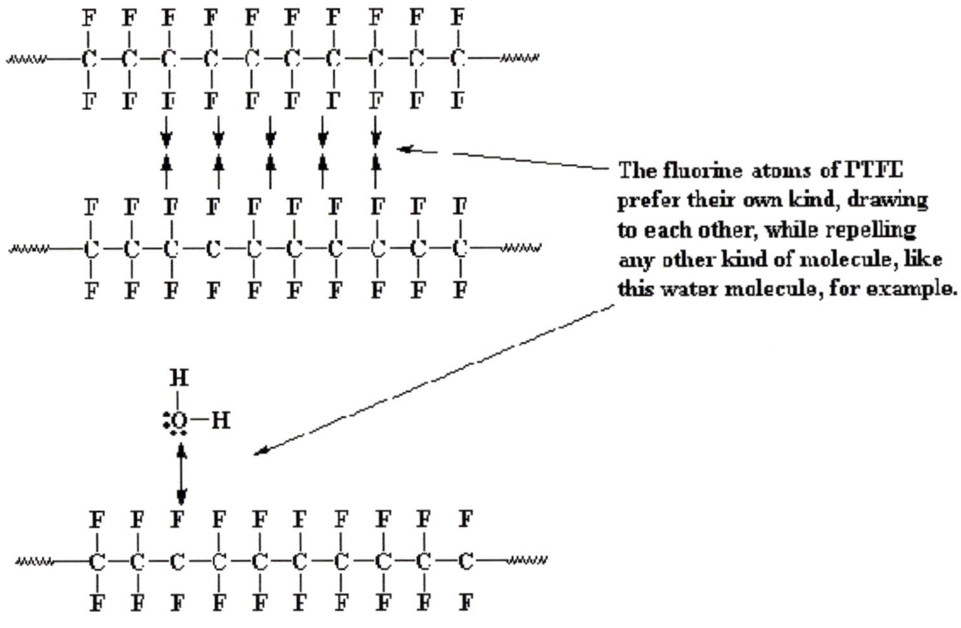

Figure 2. The molecular structure of PTFE

PTFE repels everything and hence no molecules can stick to the PTFE surface which makes it slippery (Fig.2). Ice is the only material that is slicker than PTFE. A thin PTFE coating over metal cooking pans makes them nonsticky with food items. PTFE can withstand a wide range of temperature (-184°C to 260°C) and is used in cold as well hot environments. It is hydrophobic (water repellant) and hence is resistant to weathering. It has fantastic chemical resistance and superb electrical insulation properties. It is the only plastic which can withstand temperatures up to 300°C. On heating to temperatures above 400°C, PTFE disintegrates with the production of carbon. Above 500°C, when heated in air, PTFE disappears altogether due to the production and escape of carbon and fluorine in the form of CO_2 and fluorine gases into the atmosphere [1]. PTFE is insoluble in common solvents and is resistant to nearly all acidic and alkaline

http://dx.doi.org/10.5772/59599

materials. It has a high dielectric strength and low dielectric loss. Due to high melt viscosity, injection or blow molding is not possible with PTFE. Only hot sintering or ram extrusion manufacturing processes which are relatively expensive are being followed for making PTFE products. In rapidly growing economies like China, the demand for PTFE has grown 5 times over the past 5 years.

2. Filler grade PTFE

PTFE undergoes creep (deformation under loading) which can be reduced with the help of high shear modulus fillers such as glass. Fillers hinder the relative movement of the PTFE molecules past one another and in this way reduce creep or deformation of the parts, reduce the wear rate of parts used in dynamic applications, and reduce the coefficient of thermal expansion. Other popular fillers used along with PTFE include carbon (improved thermal conductivity and low deformation under load), graphite (improved lubrication), bronze or stainless steel (excellent wear resistance) etc. Since PTFE powder is hydrophobic (it floats in water as seen in Fig.3) and does not flow freely, mixing it will free flowing fillers is a major task. One has to use a cryogenic medium such as liquid nitrogen to remove the electrostatic forces that hold the PTFE powder together. This technique is being used to manufacture thermoluminescent material filled PTFE discs (1:3 weight ratio) which are used for personnel radiation monitoring in India and elsewhere [2]. After radiation exposures, these discs are usually heated to 300°C during luminescence measurements and PTFE is the suitable binder for such applications. Organic liquids such as ethanol can also be used to mix free flowing fillers with non free flow PTFE since they wet PTFE powder unlike water medium. Alternately, one could use mechanical shearing force to separate the PTFE particles. The last choice is industrially viable and hence was adopted by us for manufacturing filler grade PTFE powders (Figs.4 and 5). An overview of different fillers used along with end use can be had from the brochures supplied by Dupont and other PTFE manufacturers.

Figure 3. PTFE is hydrophobic – water repellant

Figure 4. A blue pigment (3%) mixed non free flow PTFE powder.

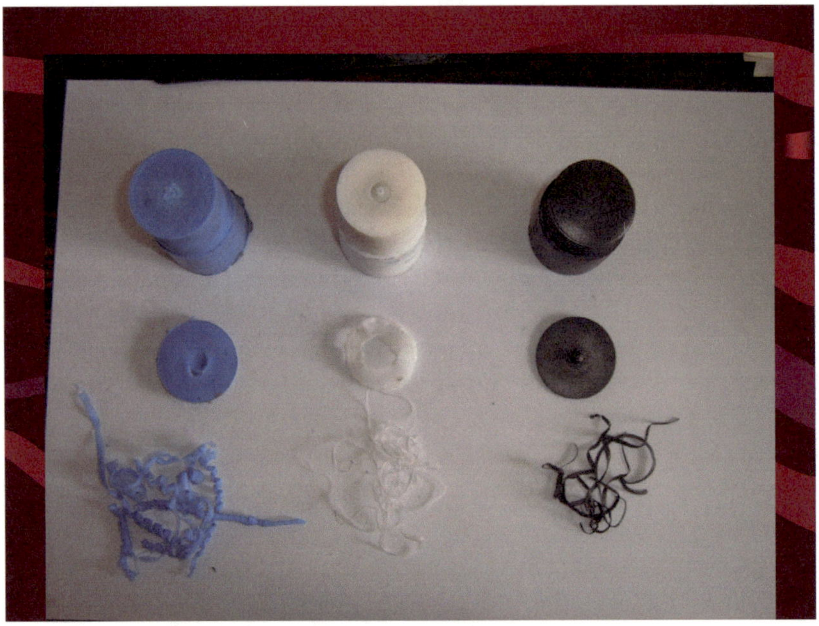

Figure 5. Sintered rods made from blue pigmented PTFE, glass mixed PTFE and carbon mixed PTFE

3. Recycling PTFE scrap material

Use of plastics has the capability to cause danger to both environment and human life. PTFE scrap can be neither incinerated nor dumped! In the waste incineration processes which have been usual up to now, highly-corrosive vapors are released which also damage the incineration

http://dx.doi.org/10.5772/59599

plant itself. Dumping of waste will in future be restricted due to nationwide regulations. In order to avoid this, Environmental Protection Agencies have taken up an action to recycle all the PTFE materials.

Recycling PTFE involves two different processes with two different end results. One involves irradiation of the PTFE scrap to heavy dose of ionizing irradiation which will reduce its molecular weight. On pulverizing, the irradiated PTFE scrap turns into *micro powder* which finds certain applications. The second method involves pulverizing the PTFE scrap without irradiation so that it becomes reusable like virgin PTFE itself.

4. Irradiation stability of PTFE

From Table 1 it is seen that among all plastics, PTFE has the least stability against ionizing radiation. Heavy gamma or electron irradiation (several kilo Gray) has been found to break down carbon-carbon bonds in the polymer chain in the PTFE scrap and reduce its molecular weight which makes it very brittle and the end product is a white, free-flowing PTFE powder which was found to be useful as additives in other materials or systems (see Fig.6). While the turnings of PTFE scrap before irradiation are tough and elastic, those after irradiation in air crumbles into a powdery material. The molecular weight of irradiated PTFE is in the range of a few tens of thousands to a few hundreds of thousands, compared to several million for the unirradiated resins. When irradiated in vacuum or inert atmosphere, the cleavage of the bonds produces highly stable radicals. The recombination of these stable radicals prevents rapid degradation of PTFE, as the molecular weight rebuilds. When irradiation is conducted in air, as is the case in the present experiment, the radicals react with oxygen leading to smaller molecular weight PTFE chains fairly quickly.

Figure 6. PTFE scrap before (left) and after gamma irradiation (right)

Plastic	Ionizing Radiation Stability
ABS	Fair
Amides	
– Aliphatic	Fair
– Aromatic	Excellent
Cellulosics	Fair
Fluoroplastics	
– PTFE	Poor
– Polychlorotrifluoroethy lene	Fair
– PVF	Good
– Polyvinylidene fluoride	Good
– Copolymers of ethylene & TFE	Good
– Polycarbonate	Good
Polyesters – aromatic	Good
Polyolefins	
– PE	Good
– PP	Fair
– Polymethylpentene	Good
– Copolymers	Good
Polystyrene	Excellent
Polystyrene acrylonitrile	Good
Polysulfones	Excellent
Polyvinyls	
– PVC	Good
– Copolymers	Fair

Table 1. Radiation Stability of thermoplastic polymers [3]

4.1. PTFE Micropowder

The irradiated and pulverized PTFE scrap differs from PTFE granular resins and fine powders because of the very small particle size, typically in the range of 2 to 20 μm (and hence the word *micropowder*), low molecular weight and the way they are handled and processed. Micro powders were first developed as an outlet for the disposal of scrap resin. The recycled PTFE powder cannot be used for the same applications as PTFE since its properties are quite different; it cannot be molded in the same way and it does not possesses the same plastic properties of virgin PTFE. However, it can be used as an additive/lubricant in other materials (printing ink, thermoplastics, elastomers, coatings and lubricants) and can be used over a wide range of temperatures from -190 to 250°C and depending on the application, may provide non-stick properties, improved lubricity, better wear resistance and reinforcing properties [4]. More of these applications can be seen in the pamphlets published by Dupont. Micro powders down to four micron range used as additives for Inks, Oils, Lubricants, Paints and Coatings, Cosmetics and Thermoplastics show enhanced lubrication and wear resistant properties.

Irradiation time in sec	Dose, Mrad	Average particle size (μm)
2.5	5	11.1
5.0	10	5.3
7.5	15	2.5
10	20	1.5
12.5	25	0.9

Table 2. Effect of electron irradiation dose on PTFE Micropowder Particle Size

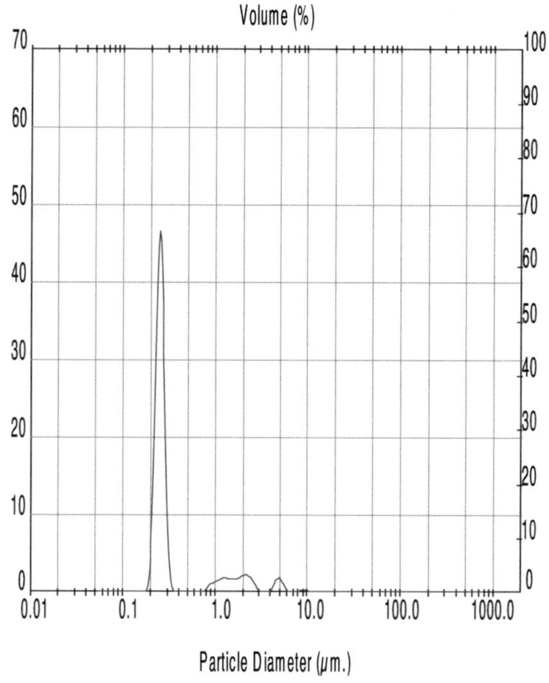

Figure 7. Particle size analysis of micro powder using laser light scattering.

Table 2 shows that with increasing electron irradiation dose in the range 5 – 25 Mrad, the average particle size of the PTFE micropowder decreases from 11.1 to 0.9 μm. The melt flow index goes up as the molecular weight of the powder goes down with increasing dose of irradiation. Electron irradiation is reported to cause cleavage of bonds and generation of gases such as HF acid vapor, which must be removed by means of adequate ventilation from the processing areas. Electron irradiation also increases the temperature of the sample which is held below 121°C by fractionating the irradiation. Both these problems are much less severe in gamma irradiation as the irradiation rate is several orders of magnitude less than that of electron irradiator.

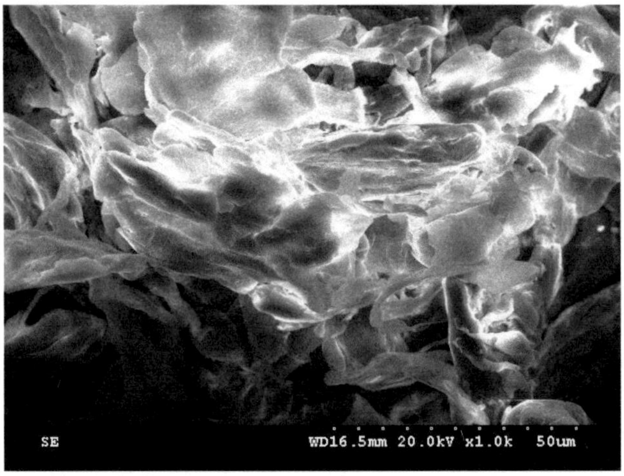

Figure 8. Scanning electron micrograph of PTFE micro powder - irradiated and pulverized

Particle size distribution of the micropowder was carried out using a laser light scattering instrument (Aerosol Dust Monitor Model 1.108 of M/S GRIMM Aerosoltechnik, GmbH, Germany). The scattering angle by a single particle is inversely proportional to the size of the particle. By measuring the forward angle of scattering and intensity of the scattering light, both size distribution and number concentration could be obtained. Fig.7 shows that an average particle size of 0.26 μm was obtained on electron irradiation of the PTFE scrap followed by milling which was achieved by a jet mill or a hammer mill. In the jet mil particles strike against each other, causing them to fracture into smaller particles. The flaky morphology of imported scrap powder seen in Fig.8 offers better lubricant property of the micro powder.

5. Reprocessing of unirradiated PTFE scrap

Unirradiated PTFE scrap can be recycled into many other products and used for rods, tubing and sheets etc by pulverization followed by suitable heat treatments similar to that of virgin PTFE. The recycled PTFE is known technically as "Reprocessed" or "Repro" PTFE or mechanical grade PTFE. Though off-white in color reprocessed PTFE has certain advantages over virgin PTFE namely low creep and better mechanical strength.

Unlike other thermosetting plastics, on heating to high temperatures PTFE products do not melt but only soften above 327°C. Therefore techniques other than those used for recycling conventional thermosetting plastics such as polyethylene have to be employed for recycling PTFE. One way is to grind PTFE scrap into fine powder and sinter it under pressure after suitable heat treatment and cleaning process. Another way is to decompose PTFE into its raw gaseous components (tetrafluoroethene and hexafluoropropene), cleaned and fed back into the production of new PTFE. Dyneon GmbH is building a pilot plant in Burgkirchen, Germany based on the latter technique, to recycle PTFE scrap. It will have capacity to recycle 500 metric tons of PTFE waste annually.

The most common way is to blend the pulverized scrap fine powder with pure PTFE to be used either in compression molding or ram extrusion. Before grinding, the scrap is usually shredded (Figs. 9 and 10) and heated to above its melting point to remove any volatile contaminants. Once ground, it is treated with acid to dissolve inorganics after which it is washed

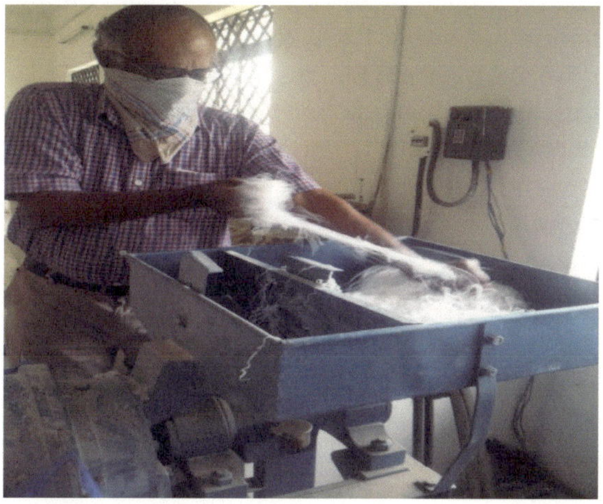

Figure 9. PTFE scrap being shredded

Figure 10. Shredded scrap

Reprocessed PTFE grade powder is manufactured from pre-sintered PTFE shavings, scrap, etc. It exhibits most of the properties that the virgin grade does but is subject to occasional contamination within the material. This is the grade of choice when cost is a major concern and cleanliness is not an issue. When repro grade is mixed with PTFE the cost comes down. Such mixed grades are used when high purity is not required such as non critical chemical, electrical and mechanical applications [5-8]. However, virgin grade PTFE is the material of choice for use in pharmaceutical, food and beverage, and cosmetics industries or for medical/ electrical applications. Virgin PTFE has better friction characteristics, which may be important in some applications. Reprocessed grade PTFE is produced for thin sheets with a maximum thickness of 0.250", For thicker sheets virgin PTFE is used. However, virgin PTFE is known to undergo creep – deformation under load whereas the compressive strength and deformation under load for reprocessed PTFE are superior to virgin PTFE. Reprocessed grade PTFE also has superior wear resistance than virgin PTFE. Reprocessed grade PTFE rods are available in diameters ranging from 1/8 to 4 inch and lengths of 6 to 12 ft. Reprocessed PTFE is frequently specified for high performance bearings and bushings, particularly in applications that require resistance to corrosive chemicals.

5.1. PTFE has different grades

Grade A: 100% virgin material.

Grade B: 70% virgin material, with 30% recycled material.

Grade C: 50% - 50%

Grade D: 30% virgin, 70% recycled.

Grade E: 100% recycled

High purity reprocessed PTFE is white in color similar to virgin PTFE and is used for applications ranging from extruded billets or molding into tubes, gaskets and ball-valve seats. Lower grade off-white reprocessed PTFE is blended with pre PTFE and is used for packing materials for valve stems and other applications (Table 3).

Repro (%)	**< 5%**	**10-20%**	**>20%**
Visual Slight off white	No notable change		Off white
Component finish	Smooth	Rough	Rough – with powdery burrs
Water absorption	< 1%	> 1%	> 2%
Chemical resistance	No notable change		
Tensile strength	Slightly reduced	Reduced by 10%	Reduced by 20%
Dielectric strength	Slightly reduced	Reduced by 10%	Not suitable
Wear resistance	Slightly reduced	Slightly reduced	Not suitable

Table 3. Properties of different grades of reprocessed PTFE powder.

5.2. Issues to be tackled in manufacturing reprocessed PTFE powder are the following

i. Difficulties in grinding the scrap into a fine powder

ii. Discoloration due to burning of volatiles and organic materials in ground powder

iii. Unlike virgin PTFE scrap filled PTFE or scrap powder could not be sintered at high temperatures (350 -400°C) at atmospheric pressure

iv. Agglomeration and Sintering of the scrap powder during thermal anneal caused cracks in reprocessed PTFE billets during sintering at atmospheric pressure

Efforts needed to solve the above problems are described below.

Figure 11. Charred organic impurities on pre-heating the shredded scrap

Figure 12. Major charred impurities are removed by handpicking

Figure 13. Shredded PTFE scrap after pre-heating treatment

5.3. Grinding PTFE scrap - Kirk-othemer encyclopedia of chemicals

A technique known as Shear Extrusion Pulverisation based on Bridgeman – Anvil was used for this purpose. This technique is also known as

- Double disc mill
- Solid state pulverization (SSP)
- Pressure shear pulverization (PSP).

It is a physico-chemical process in which cohesive forces within the polymer are broken by means of mechanically induced stress.

The process is based on "Bridgeman phenomenon" and is realized inside a specially deigned pulverizer

In PSP, the polymer is subjected to simultaneous action of axial compression and shear stress between two mirror-like smooth working surfaces in the pulverizer and the pulverized

The process is realized below the melting point of polymers so the technical properties of recycled polymers remain unchanged after polymerization

Figure 14. Pre-sintered scrap powder

Figure 15. Ground scrap powder after pre-sintering

Figure 16. Ground PTFE powder being mixed with virgin PTFE powder.

Figure 17. Mixture being cold pressed into pellet.

Figure 18. Fillers like graphite or pigments normally do not melt or agglomerate nor interact with PTFE during sintering at atmospheric pressure but this is not the case with reprocessed PTFE when used alone or used as a filler with virgin PTFE

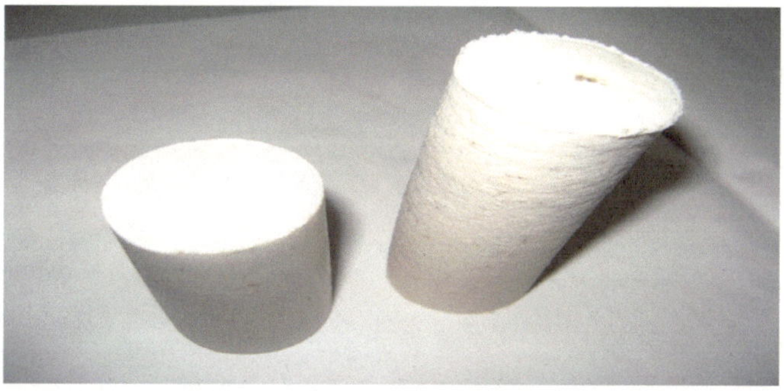

Figure 19. Reprocessed PTFE filled (40%) PTFE pellet before (left) and after (right) sintering at atmospheric pressure. The surface became rough on sintering due to agglomeration of reprocessed particles and their migration to surface of the pellet. Reprocessed PTFE is amenable to sintering only under pressure or under ram extrusion.

Figure 20. Sintered rods at atmospheric pressure – note pellets from 100% virgin scrap (second from left) crack and pop out perhaps due to the release of volatile gases whereas the pellet from 100% virgin PTFE (extreme right) are milky white in color and exhibit good integrity on sintering at atmospheric pressure.

6. Compression molding

Compression molding is a method of molding in which the molding material, generally preheated, is first placed in an open, heated mold cavity. The mold is closed with a top force or plug member, pressure is applied to force the material into contact with all mold areas, while heat and pressure are maintained until the molding material has cured. Its advantage lies in its ability to mold large, fairly intricate parts.

Initially it was thought that the die and plunger could be heated during compression in the hydraulic press. A strip heater was wound around the die to heat the die with the help of a heating programmer and a thermocouple was inserted into a hole made specifically for this purpose in the die to measure the temperature during compression, The experimental arrangement is shown below.

Figure 21. Compression molding

The strip heater wound around the die, however, could not heat it to the desired temperature even after several hours of operation as the heat capacity of the die made out of hardened steel was much high and so this attempt was discontinued. Instead, after cold pressing in the press, the die and plunger itself with the PTFE disc inside it was kept inside the high capacity air over under mechanical pressure.

Figure 22. Cold pressed pellet along with the die and plunger being pressurized with the help of a C-clamp

Figure 23. Die under C-clamp kept inside the air over for sintering near 400°C

Figure 24. Repro filled PTFE Discs when sintered inside the die and plunger under C-clamp show discoloration which runs through its volume although good surface smoothness is seen and cracks disappeared totally. This was attributed to carbon production on reaction of volatiles with the die material. Even 100% virgin PTFE pellet show slight discoloration but only on its surface (both sides) when sintered under C clamp

Figure 25. 10% mixture, mixee mixed, 300 deg C, 2h pre-heated, 2500 psi, No pressure during sintering at 400 deg C in a furnace. Sample is white in color but surface is rough.

The reason for the discoloration could be due to carbon generation on reaction of volatiles with the stainless steel die under pressure. This can be avoided if a pathway can be provided for the escape of volatile gases during sintering under pressure. This will need fabrication of a new die and plunger with a series of holes. While high pressure (2000 to 3000 psi) is required during cold pressing powder into pellet, a relatively lower pressure (500 to 1000 psi) should suffice during sintering. Further efforts were made by reducing the die pressure to 2500 psi and sintering without pressure.

This showed that mild pressure during sintering is a must. High pressures with clamp not only discolor the pellets but also fuse them with the die. So a compromise in pressure during cold pressing as well as provision to let out the volatile gases are necessary

6.1. Sintering treatment

Sintering temperatures were varied from 350 to 450 deg C and duration from 15 min to 1h. From the points of view of polish, smoothness and strength, the best treatment was found to be 380-400 deg C, 1h which is the same used for sintering virgin PTFE. Lower temperatures resulted in poor strength due to under-sintering while higher temperatures resulted in poor strength as it reduced the polymer strength.

7. Ram extrusion

Ram extrusion enables continuous processing of PTFE [9,10]. The PTFE powder (virgin or reprocessed) is fed into a cylindrical extrusion pipe hydraulically while at the same time compressing it by means of a ram and transported through the pipe, which is heated up to sintering temperature in the range 370 to 400°C (Fig.26). The ram is then withdrawn, the die tube re-charged with powder and the cycle repeated. This way the powder is continuously fed into the heating section of the die tube where it is sintered and then it passes through a cooler section from which the finished products (rods, tubes etc) flow out continuously which are cut into desired lengths. Apart from PTFE materials like ultra high molecular weight polyethylene as well as their compounds can be ram extruded. Uniform distribution of powder into the cavity is essential. The powder should exhibit good flow properties. The extrusion pressure is in the range of 26 to 74 MPa depending on the length of the cold zone above the heated section of the die tube and the extrusion rate is 3 m/h. The heated length of the tube can vary from 44 to 90 cm.

The 50% recycled PTFE ram extruded PTFE rod made in China is shown in Fig.27. It is clear that the discoloration in ram extruded rods (top) made by us was due to the production of carbon (Fig.28). This conclusion was confirmed when the discoloration reduced on subsequent anneal at 400° C for 1h in air oven which led to the escape of carbon into atmosphere in the form of carbon di oxide gas. As a result the extruded rod (bottom) exhibited biscuit color all thorough its volume without any black patches. This is a significant result and confirms that the discoloration is caused by carbon production. The fluorine being a gas produced on disintegration of fluorocarbon perhaps has already escaped during the sintering while carbon in the absence of oxygen has deposited within the rod. Acid treatment before sintering has been found to improve the color further.

Figure 26. A typical ram extruder with 100% virgin extruded PTFE rod.

[Zhejiang, China (Mainland)]

Figure 27. 50% recycled PTFE rod from Deqing VRT Plastic Industrial Co., Ltd.

Figure 28. A 40% repro + 60% virgin PTFE mixture was ram extruded into long rods of diameter 17 mm with the help of a local industry in India. The rods came out fine from the points of polish and strength but was dark in color with patches all through the length and volume due to carbon deposition.

8. Conclusions

Among all plastics, PTFE has the least stability against ionizing radiation, a property which is used to break down carbon-carbon bonds in the polymer chain in the PTFE scrap and reduce its molecular weight which makes it very brittle and the end product is a white, free-flowing PTFE powder which was found to be useful as lubricant additive in other materials or system such as printing ink, thermoplastics, elastomers, coatings and other lubricants. While the turnings of PTFE scrap before irradiation are tough and elastic, those after irradiation in air crumbles into a powdery material. The molecular weight of irradiated PTFE is in the range of a few tens of thousands to a few hundreds of thousands, compared to several million for the unirradiated resins.

Unirradiated PTFE scrap could be successfully ground with the help of a commercial shredder and milling machine. A technique based on Shear Extrusion Pulverization based on Bridgeman – Anvil was successfully used to grind PTFE scrap into a fine powder. Suitable pre-heat treatments were arrived at to remove organic and other volatile impurities. Since normal sintering procedures used for molding virgin PTFE did not work with repro filled PTFE, sintering under pressure and ram extrusion techniques were tried to mold them. Repro filled PTFE Discs up to 40% tried so far when kept inside the die and plunger and pressurized with a C-clamp show discoloration which runs through its volume although good surface smoothness is seen and cracks disappeared totally. This was attributed to carbon production on reaction of volatiles with the die material. The discoloration can, however, be avoided if a pathway can be provided for the escape of volatile gases during sintering under pressure That will need fabrication of a new die and plunger with a series of holes. While high pressure (2000

http://dx.doi.org/10.5772/59599

to 3000 psi) and hence a solid die and plunger are required during cold pressing powder into pellet, a relatively lower pressure (500 psi) should suffice during sintering.

A 40% mixture was successfully ram extruded by us into long rods of diameter 17 mm. The rods came out satisfactorily from the points of polish and strength but was dark in color with patches all through the length and volume due to carbon deposition. Due to escape of carbon, the discoloration of ram extruded rod considerably reduced on annealing in air at 400°C in atmospheric pressure as the volatile gases found an easy pathway into atmosphere. This is a significant result and confirms our view that the discoloration is caused by carbon production. The fluorine being a gas produced on disintegration of fluorocarbon perhaps has already escaped during the sintering while carbon in the absence of oxygen has deposited within the rod. Discoloration could be further reduced with acid treatment so off-white recycled PTFE rods could be made with ram extrusion.

Acknowledgements

This work was carried out under the TePP-DSIR (New Delhi, India) Techno Entrepreneurship funded project entitled "Development of Filler Grade PTFE powders and Recycling PTFE Scrap Materials". We thank Hindustan Nylons, India for making Ram extruded PTFE rods.

Author details

Arunachalam Lakshmanan[1*] and S.K. Chakraborty[2]

*Address all correspondence to: arunachalamlakshmanan@yahoo.com

1 Saveetha Engineering College, Thandalam, Chennai

2 Department of Scientific and Industrial Research, Techno Entrepreneurship, New Delhi, India

References

[1] AR. Lakshmanan and U. Madhusoodanan. Behaviour of $CaSO_4$:Dy Embedded Teflon Discs at high Annealing Temperatures. *Rad. Meas.* 29, 327-330 (1998)

[2] K.G. Vohra, R.C. Bhatt, Bhuwan Chandra, A.S. Pradhan, AR. Lakshmanan and S.S. Shastry. A personnel Dosimeter TLD Badge Based on $CaSO_4$:Dy Teflon TLD Discs. Health Phys. 38, 193-197 (1980)

[3] Plastics for Medical Use. Brochure No. 3 issued by Radiation Plant for Sterilization of Medical Products (ISOMED), Trombay, Bombay 400085, India.

[4] Micropowders, Dyneon™*PTFE, http://www/3m.com/product/m_index/Micropowders,_Dyneon(TM)_PTFE_(BW)_00.jhtml

[5] Repro-Lon Texeas – PTFE Resins. http://www.reprolontexas.com/products.htm

[6] PTFE Recycling Market, September 22, 2012 in PTFE, FEP, PFA. Recycling by Miller-Stephenson Chemical from Rubber World, Vol. 1, 1999. www.polyfluoroltd.com

[7] Can reprocessed resin do the job? Feb 11, 1999. Machine Design, 2010, Penton Media Inc. Edited by David S. Hotter, Sina Ebnesajjad, Vadium Lishinsky, Technical Service – Teflon Dupont Fluoroproducts, Welimington, De, USA

[8] PTFE and the "Repro" Conundrum. The definitive blog on fluoropolymers from PolyFluoro Ltd, India. www.polyfluoroltd.com Posted on May 20, 2011

[9] ASTM D4894-07 (2012) Standard Specification for PTFE Granular Molding and Ram Extrusion Materials Active Standard ASTM D4894 Developed by Subcommittee D 20.15 Book on Standards Volume.08.02

[10] Fluon (Asahi Glass Company Trade Mark). The extension of PTFE granular powders. Technical Service Note F2. Edition, AGFP, September 2002